Water and Environment for Sustainability

Naeem Shahzad
Editor

Water and Environment for Sustainability

Case Studies from Developing Countries

Springer

Editor
Naeem Shahzad
National University of Sciences
and Technology
Islamabad, Pakistan

ISBN 978-3-031-27279-0 ISBN 978-3-031-27280-6 (eBook)
https://doi.org/10.1007/978-3-031-27280-6

This Springer imprint is published by the registered company Springer Nature Switzerland AG
The registered company address is: Gewerbestrasse 11, 6330 Cham, Switzerland

Preface

This book is being published as an outcome of the 2nd International Conference on Water, Energy, and Environment for Sustainability (IC-WEES) 2022. With the growing global concerns about environmental degradation, depletion of freshwater resources, and climate change-induced disasters, the conference was focused on climate change, water, environment, and disaster risk reduction and their interrelationship with each other. Major theme revolved around Climate, Environment and Water Nexus. Owing to the papers presented in the Conference the book title was further focused toward water and environment for sustainability.

A multidisciplinary understanding of water resources and environmental concerns is provided by water and environmental management to ensure sustainable development making the understanding of water resources and environmental concerns imperative. Despite considerable advancements in the science of water and environment and their close nexus over the past century to ensure sustainability, there are still a number of problems that the scientific community has to address.

Water bodies offer a variety of beneficial services to humanity owing to their environmental, cultural, and commercial significance as ecosystems. Owing to lack of check and balances and implementation issues, environmental degradation at a rapid pace further augmented by climate change, water quantity, water quality, ecosystem health, and ecosystem services are all faced with different challenges (such as global warming, changes in land use and cover, population explosion, nutrient and organic matter loading, water conservation projects, and other anthropogenic factors). This underscores the need of multidisciplinary research, accurate monitoring, and modeling of various water bodies as it is crucial for improving the management of water resources and ensuring sustainable water usage.

This book *Water and Environment for Sustainability* brings together thirteen papers on the theme of 'water resource and environmental management.' The purpose of this book is to present some of the latest research carried out in the area of water and environmental management to ensure sustainability in a dynamic situation with uncertainty.

Most Pakistani regions (including the northern glacial regions) are getting hotter with the spatial trends showing an increase of more than 0.05 °C/year in temperature and depicting an increase of 5 °C/century, as described by Amjad (2022). In this paper, the author highlights that the indication of northern glacial regions and south-western Baluchistan getting hotter and drier is quite an alarming situation for both these regions in the contexts of frequent and extreme flooding due to snowmelt in glacial regions and frequent and intense droughts in southwestern Baluchistan. The issue needs to be tackled immediately by relevant stakeholders, organizations, and departments.

Based on the water parameters obtained at several hydrological stations along the Brahmaputra River, India, the study's overall objective was to evaluate the evolution of water quality at spatial and temporal scales. According to the pollution sources listed, anthropogenic factors such as population growth, excessive consumption, overuse, chemical explosions, and subsurface mining are found to be the main causes of pollution in the catchment area Pranjal (2022). As per the findings, it is crucial to make sure that the hydrological forces trying to control runoff are understood and incorporated into the analysis in these days of increasing assessment complexity and reliance on computer data manipulation, rather than relying on more complex sets of parameters and relationships to explain runoff.

A comprehensive overview of the recent scientific research in the field of reservoir sedimentation estimation and management has been presented in the study carried out by Bilal (2022) which is very significant for flood management in downstream areas. The results of this study can serve as suggestions for the execution of operations to flush silt from reservoirs. Additionally, it is anticipated that the observations will be helpful for assessing future dam safety and danger as well as for ensuring reservoir operating sustainability through thorough sediment management leading to flood mitigation and management in the downstream areas.

The editors wish to thank all the authors of this book for contributing the high quality papers. We also like to express our gratitude to the referees for their quick and thorough reviews of the papers. Finally, we hope that the readers will find this book to be very helpful and will share our joy.

We hope this book will help us manage water resources efficiently through environmental preservation to achieve a healthy surroundings and a sustainable future for our future generations.

Finally, we would also like to thank the excellent editorial team of *Water and Environment for Sustainability* for their efforts and professional commitment.

Islamabad, Pakistan Naeem Shahzad

Contents

Hydrometeorological Extremes and Climate Profiling of Pakistan

Muhammad Amjad

1 Introduction

Quickly evolving climatic change poses one of the biggest dangers to sustainability of life on earth. Increasing greenhouse gas emissions in the exterior atmosphere due to fossil fuel combustion is deemed the reason for a continuous rise in earth's temperatures besides polluting air. The climate change-induced phenomenon of global warming is among the major causes of environmental degradation in the modern era. Gradually rising temperatures and their influence on the cryosphere are apparent in several regions around the globe. Climate change affects human life in addition to the economy by causing disturbances in the earth's climatic responses triggering natural disasters such as floods, famines, droughts, and cyclones along with several other disasters (Shahzad & Amjad, 2022).

The issue of climate change has surfaced very strongly during the past couple of decades worldwide threatening all humans irrespective of the prosperity and wealth statuses of the countries they belong to. However, developed countries suffer less due to their financial resources and built capacity. On the other hand, underdeveloped regions of the world are more susceptible to adverse impacts of climate change. Pakistan, ranked 12th among the topmost affected countries of the world due to climate change-induced hazards (Ullah, 2017), is exposed and vulnerable to several natural hazards, including cyclones, floods, droughts, intense rainfalls, and earthquakes (Salma et al., 2012). These hazards, when combined with vulnerabilities (in the form of poverty, exclusion, and poor political decisions and actions), render people more susceptible to the impacts of hazards. Agriculture sector is most vulnerable to climate change problems and changes in cropping and productivity due to weather changes will affect the poor communities of the country. The dry land areas, including arid and semi-arid regions, are also vulnerable to these changes as

M. Amjad (✉)
National University of Sciences and Technology, Islamabad, Pakistan
e-mail: amjadiqm@gmail.com

N. Shahzad (ed.), *Water and Environment for Sustainability*,
https://doi.org/10.1007/978-3-031-27280-6_1

these regions are already facing severe water shortages and high-temperature problems. The abnormal pace of climate change is expected to have extensive impacts such as stressed agricultural production, increased variability of water availability, increase in coastal erosion and seawater intrusion, and augmented frequency of disastrous hydrometeorological extremes. In addition to its standing among the topmost affected countries in the world, Pakistan's socioeconomic conditions also render it a weak and easy target for the disastrous impacts triggered by natural hazards originating from climate change.

In addition, the burning of fossil fuels to increasing amounts remains to be the backbone of industrial development for the past century where deforestation and burning of all types of fossil fuel changed the face of this world through the rapid pace of development and this, in return, came with a huge cost, i.e., "Global Warming." Extreme emission of greenhouse gases into the environment caused global temperatures to rise rapidly which have resulted in glaciers melting at a rapid pace. This has heavily affected the hydrometeorological cycle all around the world. Countries like Pakistan are now severely prone to hydrometeorological extremes, which not only have caused billions of dollars in property damage but also have taken tens of thousands of lives. Severe drought in 1998–2002 and extreme floods in 2010 and 2022 are evidence of it, yet this threat continues to increase day by day, and the trends seem to be much more frequent. To monitor and model these events, Pakistan lags in financial and technological resources. Here, weather monitoring stations are sparsely located with negligible maintenance and incorrect data collection. In addition to that, access to weather observations for different stakeholders is one of the most difficult tasks. A precise and reliable forecast and estimation of precipitation and temperature can help in improving the processing and modeling of water budgeting, water consumption, flood and droughts forecasting, and determining crop water necessities. Periodic changes in magnitude and patterns of precipitation and temperature are significant indicators of climate change. Pakistan, as mentioned earlier, lacks weather and climate monitoring resources. This posts the need for new technology to be adopted that is handy, maintenance free, and easily accessible. Model- and satellite-based sources of climate data are such modern technology tools that might be considered potential alternatives to conventional data observation using ground-based gauge stations. However, their utilization as a potential alternative to conventionally observed data needs a comprehensive investigation of their error uncertainty(Amjad et al., 2020). The 5th generation reanalysis product, namely ERA5 (Hersbach et al., 2019), from European Center for Medium-range Weather Forecasts (ECMWF), is such a model-based source of climate data retrieval whose time series accuracy assessment has been investigated in this study.

The research literature shows the imminent climate change effects being and going to be faced by Pakistan while the country lags behind the modern world in modern tool usage and, even, in properly monitoring the climate processes due to its constrained economic resources. It is pertinent to mention that very less research has been carried out to utilize the recent technological advancements in the field of hydrometeorology in Pakistan (e.g., Ali et al., 2022; Arshad et al., 2021; Jan et al., 2022; Tariq et al., 2014; Ullah et al., 2021).

This chapter is intended to focus its investigation on the following objectives:

1. Climate profiling of Pakistan and investigating its climate change zones
2. Assessment of time series accuracy of a model-based climate dataset as a potential alternative to observed data in monitoring the hydrometeorological extremes in Pakistan

The first objective of the study is aimed at indicating the climate change zones of Pakistan and, in turn, facilitating the stakeholders of water resources management while the second objective specifically intends a valuable input to disaster risk managers in Pakistan.

2 Climate of Pakistan

Pakistan lies in the coordinate box with latitude 23° 35′ N–37° 05′ N and longitude 60° 50′ E–77° 50′ E. Pakistan has a temperate climate that varies along its topography. Generally, dry and hot toward the south and along the low plains of the Indus River covering parts of Baluchistan, Sindh, Southern Punjab, and areas toward the south of Khyber Pakhtunkhwa as well. Temperature becomes progressively cooler toward the North that includes the portions of Northern Punjab, Northern Khyber Pakhtunkhwa, Gilgit-Baltistan, and Azad Kashmir, comprising of high mountainous ranges of Himalaya where the elevations of peaks climb up to and more than 8000 m above mean sea level (MSL). Pakistan has four seasons: cold dry winter from December to February, dry spring from March to April, wet monsoon from July to September, and late monsoon from October to November. Overall, the country receives very little rainfall, however, in monsoon season, it can reach up to 200 mm. Pakistan receives maximum precipitation in its Northern mountains. Due to its mountainous topography, it is very prone to flash floods. El Niño is a significant influence on climate variability in Pakistan, with anomalies in both temperature and flood frequency and impact correlated with the El Niño cycle.

Rainfall in Pakistan varies from as little as less than 10 inches to over 150 inches per annum in several parts of the country. However, this generalization should not obscure the distinct differences existing among certain locations. For example, the coastal area near the Arabia Sea is usually warm and temperate, whereas elevations of the Karakoram Range and other mountains remain snow-covered for many months of the year. Far north regions are so cold year-round that they are only reachable by world-class climbers for a few weeks in May and June of each year. Aridity is the most persistent aspect of Pakistan's climate, and its characteristics are observed during the extremes in temperatures from May to August in several parts of the country. Pakistan is located on the edge of a monsoonal system; precipitation is unreliable, and its volume is highly unpredictable. Figure 1 shows a digital elevation map (DEM) of Pakistan where different regions having elevations ranging from mean sea level to above 8000 m are depicted.

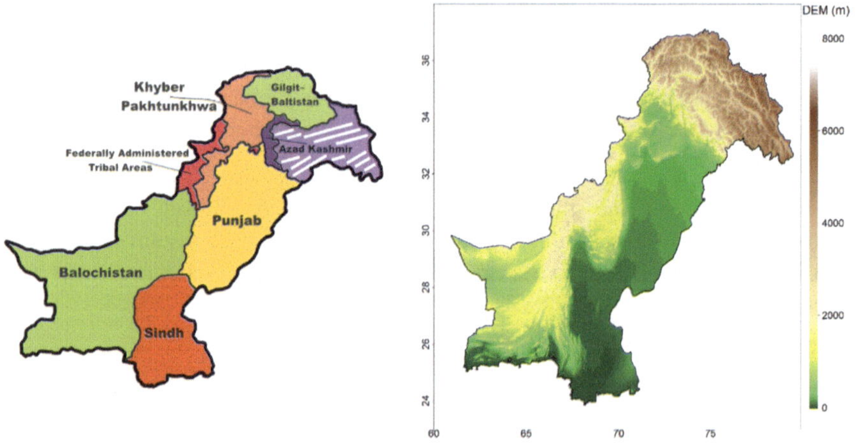

Fig. 1 Digital elevation map of Pakistan at right (Weblink for the figure at left: https://www.kindpng.com/imgv/xxomJ_administrative-map-of-pakistan-hd-png-download/)

3 Gridded Dataset for Profiling

The modern-era weather forecasting centers are releasing short-to-medium range weather forecasts that are run multiple times a day on sophisticated computational models. While producing weather forecasts, the models usually are in a continuous evolution regarding the introduction of their updated versions. There is another type of data release that is obtained by keeping the model state constant and the model is run at a relatively coarser resolution for archival data. This type of dataset is called reanalysis data. The fifth-generation reanalysis data (ERA5) of European ECMWF is based on 4D-Var data assimilation using Cycle 41r2 of the Integrated Forecasting System (IFS) of the European Center of Medium-Range Weather Forecasts (ECMWF). ERA5 benefits from a decade of developments in model physics, core dynamics, and data assimilation relative to its predecessor ERA-Interim (Hersbach et al., 2019). ERA5 features several improvements over ERA-Interim, such as a more recent model and data assimilation system, a higher horizontal resolution, assimilation of substantially more observations (including gauges and ground radars), a near-real-time release of the data, and outputs with a higher resolution ion (Beck et al., 2019). ERA5 hourly data (with a spatial resolution of 0.25°) for precipitation and temperature for the period 1979–2020 were obtained from Climate Data Store (CDS) and were converted to monthly data for further processing in this study.

4 Climate Profiling of Pakistan

Utilizing the gridded precipitation and temperature data from ERA5, climate profiling of Pakistan was carried out to show decadal variations of both the climate parameters (i.e., precipitation and temperature) spanning over four decades (i.e., 1979–2020). Historical trends were also depicted in addition to investigating the decadal variations of climatic parameters. Based on the results obtained from climate profiling, climate change zones were established indicating the whereabouts of climate change hotspots in the country.

Figure 2 shows higher precipitation values over most regions of Khyber Pakhtunkhwa. Southwestern parts of Baluchistan show dry regions (i.e., receiving very low amounts of precipitation). Central to southern regions of the country are mostly arid to semi-arid regions and receive lower precipitation. A minute decade-to-decade change between the decades is observed because of long-term averaging. A further reduction of precipitation events on a micro-scale level in the southern and southwestern parts of the country is observed in the second decade (1991–2000). The3rd decade (2001–2010) shows a reduction in precipitation over most parts of Khyber Pakhtunkhwa. Khyber Pakhtunkhwa received lower precipitation in the 3rd decade as compared to the first two decades, especially in the regions between Kohat and Bannu districts. However, the spread of precipitation amounts between 500 and 1200 mm/year is more in the 3rd decade. The color-coded yellow pixels (receiving low to high precipitation amounts) cover more area in the 4th decade as compared to the cases of the first two decades.

Pakistan records its lowest temperature values over the Himalayas (Northern mountains; Figs. 1 and 3). Punjab and Baluchistan show higher temperature values, i.e., > 20 °C. Upper Khyber Pakhtunkhwa shows mean annual temperature values ranging from 0 to 10 °C while western Punjab and central Baluchistan show average annual temperatures of < 20 °C. Regions with temperatures around 10 °C cover more areas during the 1st decade of the study period, clearly indicating the rise in mean annual temperatures in upper Baluchistan and Khyber Pakhtunkhwa. Northern parts of the country including the Himalayas show a slight increase in temperature trends over the passing decades (Fig. 3). An increasing trend in temperature is observed in the northern as well as western parts of Khyber Pakhtunkhwa province. Southern Baluchistan and Punjab, in general, show similar trends of higher temperatures. Color-coded light blue pixels (>−10 °C) cover more area in the 4th decade as compared to the first two decades.

5 Climate Trends

Climate trends in precipitation of Pakistan show mixed, but marginally inclined toward drier, trends (Fig. 4). Notably, most of Khyber Pakhtunkhwa province observes decreasing trends in average annual precipitation (Fig. 4). There are mixed

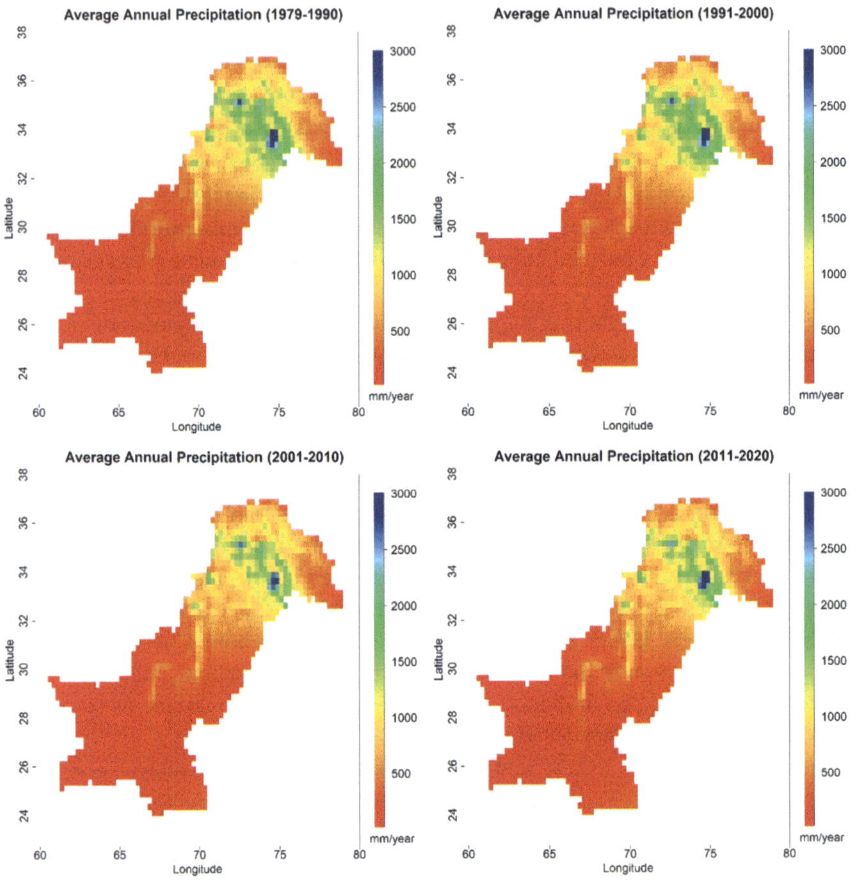

Fig. 2 Decadal annual precipitation profile of Pakistan for 1979–2020

trends over Baluchistan: both positive and negative, however, its southwestern parts have more negative precipitation trends. Few regions in the southern and south-eastern parts of the country show positive trends in precipitation. Upper Punjab and lower Khyber Pakhtunkhwa have negative trends in average annual precipitation thus implying that the agriculture in these regions is being affected due to lesser feeding from precipitation.

A rising trend in temperature by 5 °C/century (Fig. 5) in almost all of Khyber Pakhtunkhwa is alarming as the precipitation trends in this region are also negative (Fig. 4), implying that this region is getting drier and hotter. Northern glacial regions, on the other hand, also observe a rise in temperatures that can translate into higher rates of snowmelt, more intense flooding in the downstream regions, and more glacial lake outburst flooding (GLOF) events. Southwestern Baluchistan is also at risk of further rise in temperatures. Most of the agricultural regions of western Punjab are becoming hotter (Fig. 5).

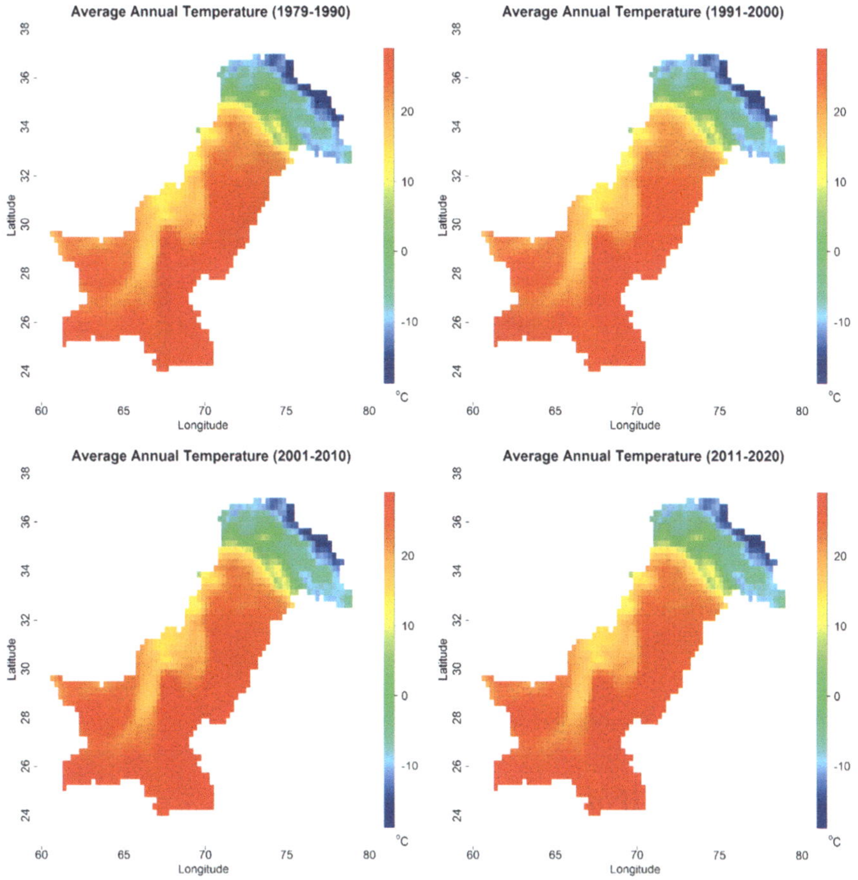

Fig. 3 Decadal annual temperature profile of Pakistan for 1979–2020

6 Climate Change Zones

Summarizing the climate profiling of Pakistan, it is obvious from the obtained results that almost all the northern regions of the country are getting drier and hotter (Fig. 6) which is an alarming situation given that these regions comprise snow-covered watersheds and glaciers that contribute to most of the feeding to rivers in Pakistan. Hotter temperatures in northern regions directly infer more frequent flash floods and GLOFs. Baluchistan is getting hotter and drier, which predicts the risk of more frequent and more intense droughts in the future. Only some agricultural regions in eastern Punjab lie in the scarce wet and cold zone (Fig. 6).

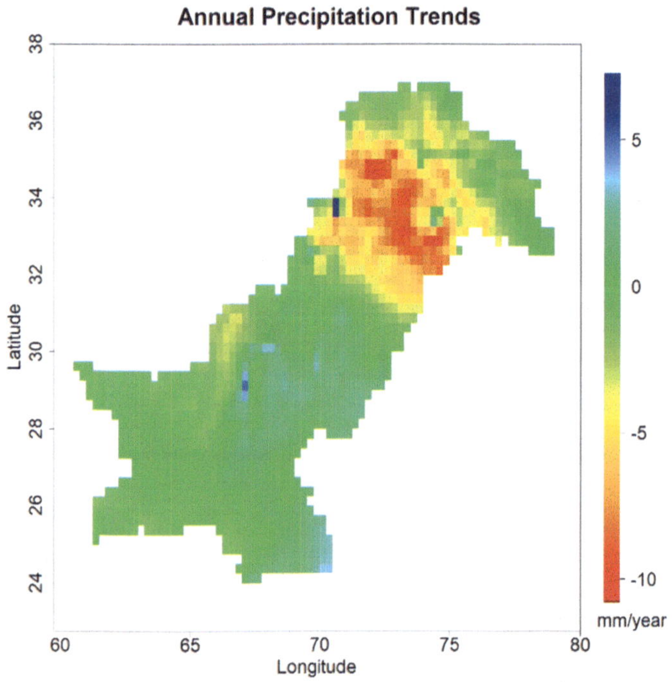

Fig. 4 Annual average precipitation trends in Pakistan for 1979–2020

7 Investigating Hydrometeorological Extremes

As mentioned in the introduction section, Pakistan is facing twofold challenges in the view of hydrometeorological extremes: the increasing frequency of extreme events and the lack of an appropriate monitoring network. The impacts of the former challenge become manifold in the presence of the latter problem. These circumstances demand investigation of potential alternatives for their accuracy in the detection and monitoring of hydrometeorological extremes. One such potential alternative could be ERA5 which is already introduced in this chapter (please refer to Sect. 3). Here, the accuracy of ERA5 for extreme events is assessed by using observed data at gauge stations (15 in total) as a reference. Locations of these gauge stations are plotted over DEM of Pakistan (Fig. 7) while their names and location information are tabulated in Table 1.

7.1 Pre-processing of Datasets

As the gauge-based observation data is point data in its nature, potential errors have resulted when interpolation techniques are applied to convert the point data into

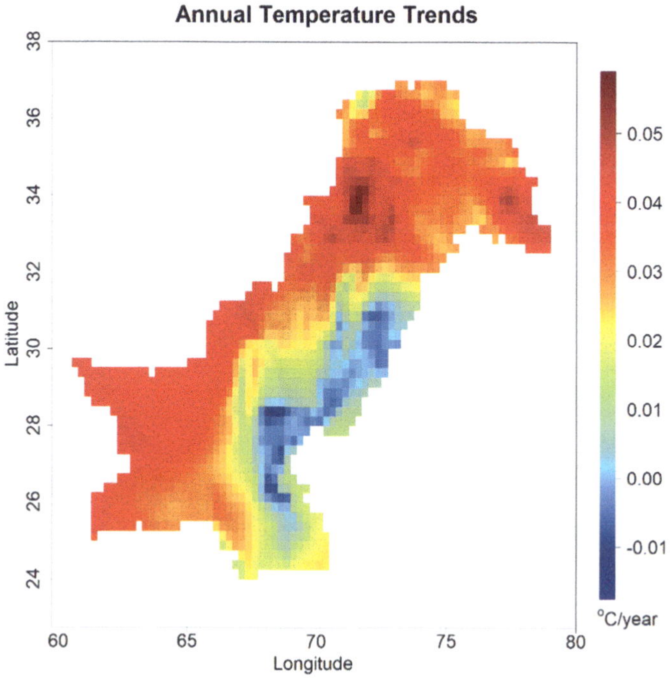

Fig. 5 Annual average temperature trends in Pakistan for 1979–2020

Fig. 6 Climate change zones of Pakistan

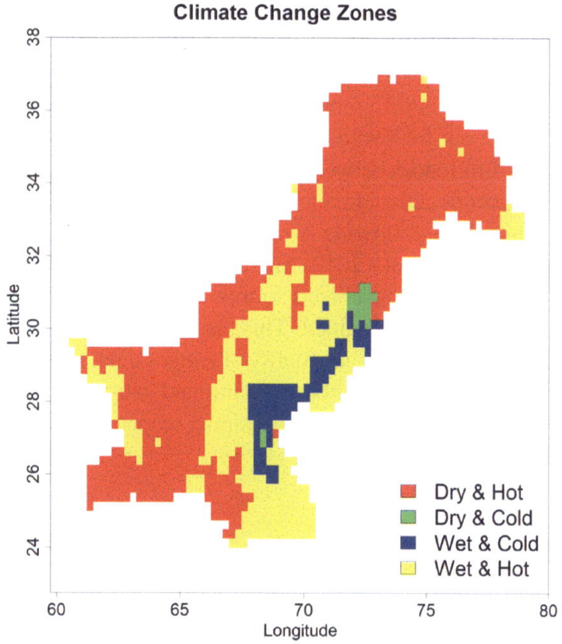

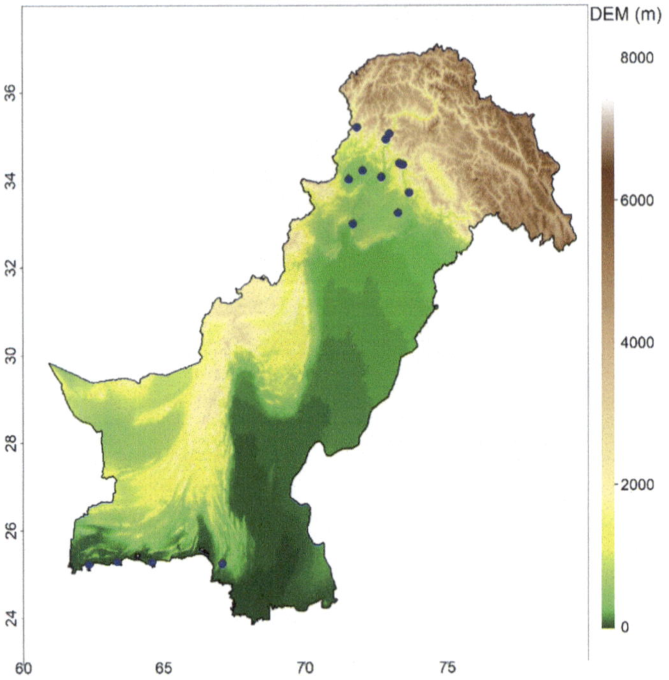

Fig. 7 Locations of 15 gauge stations plotted over DEM of Pakistan (locations shown as blue dots)

continuous spatial precipitation data (Kidd & Huffman, 2011). Moreover, the spatial representativeness of gauge-based observations (i.e., point) and model-based datasets (i.e., grids) is different. The studies performing comparisons of such datasets, in general, opt for two different methodologies commonly used to resolve the spatial scale differences between the gauges data and the gridded products: either the grids of the product data that are closest to the gauge stations are extracted (point-to-grid methodology following Amjad et al., 2020; El Kenawy et al., 2015; Heidinger et al., 2012; Islam et al., 2012) or the station-based observations within the grids of the product data are averaged so that a compatible, as well as the spatially distributed estimate, can be obtained. This study followed the first methodology, which allows for keeping the original quality of gauges data. Hence, ERA5 data was extracted over point locations of the above-mentioned gauge stations for accuracy assessment (i.e., the values of ERA5 grids containing respective 15 gauge stations were extracted).

Table 1 Information on gauge stations

Sr #	Station	Coordinates	Data Availability
1	Balakot	34° 32′ 49″ N, 73° 21′ 21″ E	2010–2019
2	Dir	35° 11′ 55″ N, 71° 52′ 19″ E	
3	Domel	35° 06′ 28″ N, 73° 00′ 07″ E	
4	Besham	34° 01′ 45″ N, 71° 31′ 24″ E	
5	Gujar Khan	25° 14′ 39″ N, 67° 06′ 47″ E	
6	Peshawar	31° 19′ 0″ N, 72° 11′ 0″ E	
7	Gwadar	25° 15′ 07″ N, 62° 17′ 07″ E	
8	Pasni	25° 16′ 19″ N, 64° 36′ 30″ E	
9	Mardan	25° 15′ 25″ N, 63° 24′ 54″ E	
10	Tarbela	34° 54′ 45″ N, 72° 52′ 04″ E	
11	Palandri	33° 36′ 12″ N, 72° 11′ 16″ E	
12	Hub dam	33° 15′ 35″ N, 73° 18′ 12″ E	
13	Massan	34° 12′ 20″ N, 72° 02′ 22″ E	
14	Ormara	33° 42′ 52″ N, 73° 41′ 15″ E	
15	Pattan	34° 07′ 35″ N, 72° 48′ 37″ E	

7.2 *Moving Averages*

The method used for evaluation is 3-monthly moving averages and the parameter considered was only precipitation as the temperature variations get dampened when averaged over 3-month time windows.

Precipitation from both the datasets (i.e., observed data at gauge stations, which will be referred to as "obs" from here on, and ERA5 dataset) was converted to 3-monthly moving averages in the form of time series, and these time series were then plotted in parallel to gather inferences on how well ERA5 captures the peaks in the observed data.

Furthermore, scatter plots were prepared by keeping moving averages of observed precipitation data at the x-axis and those from ERA5 precipitation data at the y-axis, to investigate the linear relationship between the two datasets.

Notably, three coastal stations (Gwadar, Ormara, and Pasni) have relatively flatter moving averages, resulting into having almost no troughs (i.e., droughts). Here, ERA5 followed the overall pattern as per the observed data (Fig. 8). Peshawar and Pattan have some missing observed data, the evaluation of ERA5 was done by using fewer extreme events over these two stations. Figure 8 also indicates that ERA5 followed the normal variation but underestimated the extreme precipitation. The time series for Palandri shows a consecutive series of high-peak flooding events during the start of 2014 toward mid-2016; ERA5 tends to capture the pattern of these events but it again largely underestimates the peaks. ERA5 captured the fluctuation pattern of observed data very well at Mardan, Tarbela, and Gujjar Khan which might be associated with a major reason being the flatter topography of these cities. Overall, ERA5 did follow the fluctuation pattern of most of the observed precipitation extremes.

Over some stations like Palandri, ERA5 displayed a lower correlation (R: 0.75) compared to most of the other stations (Fig. 9). ERA5 performed well in the case of Peshawar (R: 0.81), Mardan (R: 0.85), Tarbela (R: 0.88), and Gujjar Khan (R: 0.84). It is again pertinent to notice here that these three stations are located on relatively flatter topography resulting in better trend-capturing by ERA5. ERA5 also performed better in linear trend-capturing of Gwadar (R: 0.85) and Pasni (R: 0.851) which again are relatively flatter surfaces. Among all stations, Dir (R: 0.89) and Tarbela (R: 0.88) stood out regarding the trend-capturing ability of ERA5. The performance of ERA5 in the detection of extreme events over Domel and Hub Dam seems the poorest among all the stations.

7.3 *Inferences to Floods and Droughts*

Capturing the hydrometeorological extremes, triggered by weather processes that themselves are chaotic in nature, is considered a hard task for model-based weather and climate estimation sources. However, motivated by recent advancements in the subject fields of forecasting and reanalysis, this study intended to put an effort behind the accuracy assessment of model-based reanalysis (ERA5) in capturing the extremes in Pakistan.

The peaks in the 3-monthly moving averages time series generally indicate the occurrences of floods (high or positive peak) or droughts (low or negative peak). In addition, the impact of a positive or negative peak amplifies when it is followed by another peak of the same orientation. Overall, ERA5 was witnessed to match the peaks in the observed time series very well where the observed negative peaks were captured by ERA5 more consistently as compared to positive ones. Although there were instances when ERA5 underestimated the exact magnitude of certain peaks, yet its efficiency in capturing the extreme patterns is quite admirable given that it is a model-based reanalysis product having moderate spatial resolution.

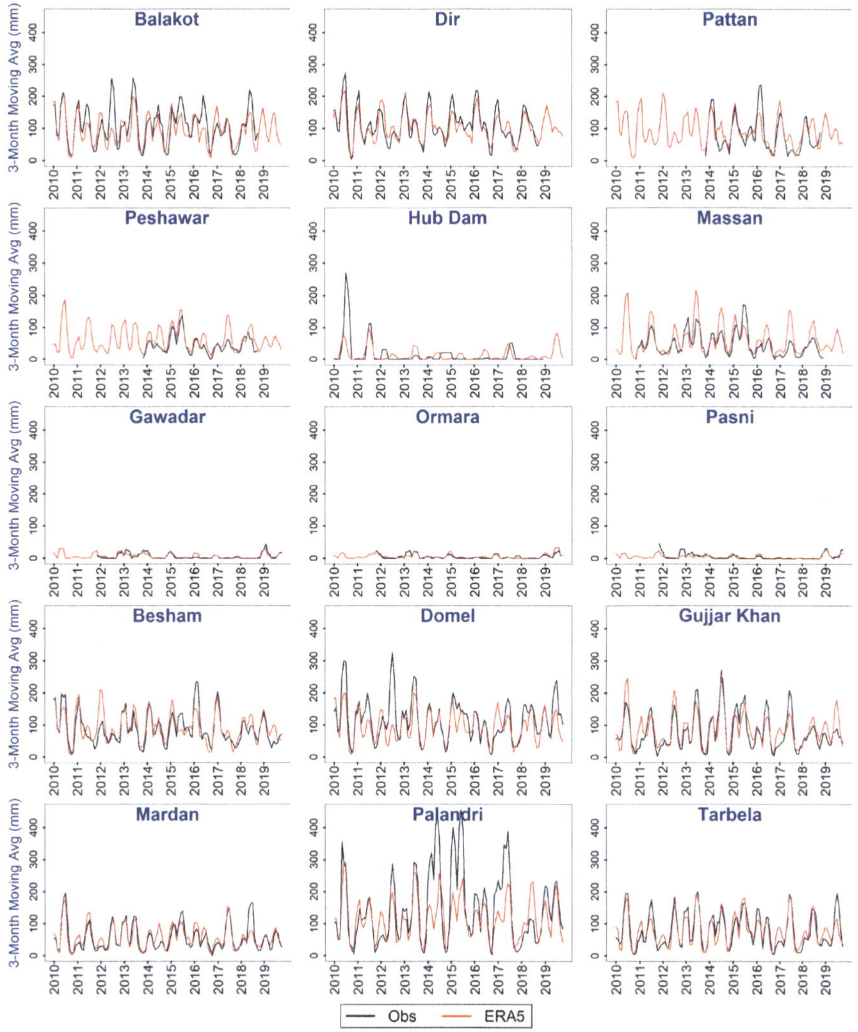

Fig. 8 Time series of 3-monthly moving averages of observed and ERA5 precipitation data sets over 15 gauge stations

8 Conclusions

The study focuses on climate profiling and the hydrometeorological extreme investigation of Pakistan which is among the worst-affected countries in the world. Climate profiling and trend zoning for precipitation and temperature was conducted by using the gridded climate reanalysis dataset ERA5 from ECMWF while the same dataset

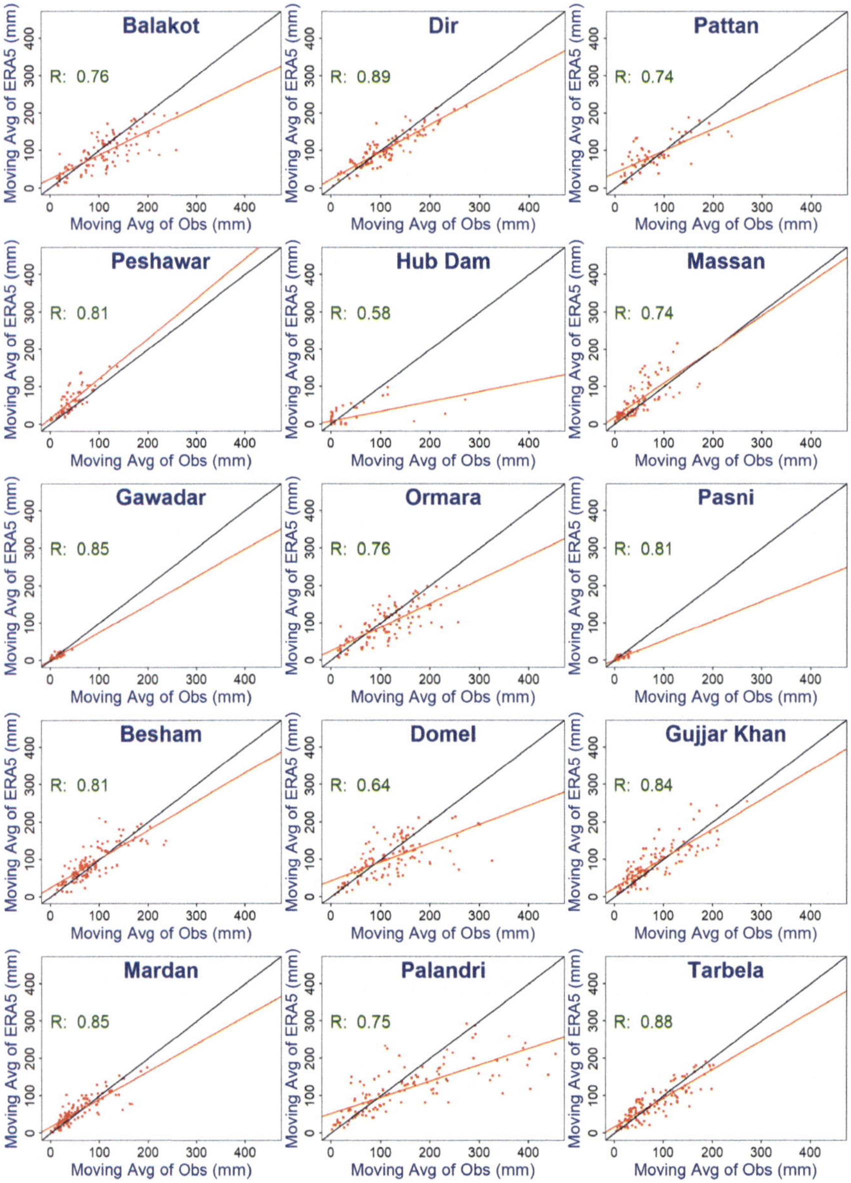

Fig. 9 Scatterplots for 3-monthly moving averages for ERA5 and observed precipitation over 15 gauge stations

was investigated, afterward, for its accuracy in the estimation of hydrometeorological extremes; the investigation was conducted by taking ground-based gauge stations data as truth.

The spatial plotting of ERA5 precipitation trends concludes that the annual precipitation amounts received by most parts of Khyber Pakhtunkhwa have a strong decreasing trend of up to 10 mm/year which is quite an alarming situation as this region is also getting increasingly drier. In addition to it, most regions (including the northern glacial regions) of Pakistan are getting hotter with the spatial trends showing an increase in temperature of above 0.05 °C/year which depicts an increase of 5 °C/century. Most regions of Pakistan fall in the climate change zone of "dry and hot" inferring that they show trends of getting drier and hotter. Indication of northern glacial regions and southwestern Baluchistan getting hotter and drier is quite an alarming situation for both these regions in the contexts of frequent and extreme flooding due to snowmelt in glacial regions and frequent and intense droughts in southwestern Baluchistan.

ERA5 underestimated the extreme events causing floods and remained on the wetter side while following very low precipitation events. It largely struggled in its accuracy over regions of complex topography. However, ERA5 produced its best results on relatively flatter topography, e.g., Gujjar Khan, Peshawar, Mardan, and coastal stations, etc. Overall, ERA5 efficiently captured the fluctuation patterns of extreme events in Pakistan.

References

Ali, M. A., Hassan, M., Mehmood, M., Kazmi, D. H., Chishtie, F. A., & Shahid, I. (2022). The potential impact of climate extremes on cotton and wheat crops in Southern Punjab, Pakistan. *Sustainability, 14*(3), 1609.

Amjad, M., Yilmaz, M. T., Yucel, I., & Yilmaz, K. K. (2020). Performance evaluation of satellite- and model-based precipitation products over varying climate and complex topography. *Journal of Hydrology, 584*, 124707.

Arshad, M., Senftl, T., & Vanselow, K. A. (2021). Performance evaluation of ERA-5, JRA-55, MERRA-2, and CFS-2 reanalysis datasets, over diverse climate regions of Pakistan. *Weather and Climate Extremes, 33*, 100373.

Beck, H. E., Pan, M., & Roy, T. (2019). Daily evaluation of 26 precipitation datasets using Stage-IV gauge-radar data for the CONUS. *Hydrology and Earth System Sciences, 23*(1), 207–224. Available at: https://doi.org/10.5194/hess-23-207-2019

El Kenawy, A. M., Lopez-Moreno, J. I., McCabe, M. F., & Vicente-Serrano, S. M. (2015). Evaluation of the TMPA-3B42 precipitation product using a high-density rain gauge network over complex terrain in northeastern Iberia. *Global and Planetary Change, 133*, 188–200. Available at: https://doi.org/10.1016/j.gloplacha.2015.08.013

Heidinger, H. et al. (2012). TRMM rainfall correction over the Andean Plateau using wavelet multi-resolution analysis. *International Journal of Remote Sensing, 33*(14), 4583–4602. Available at: https://doi.org/10.1080/01431161.2011.652315

Hersbach, H. Bell, B., Berrisford, P., Horányi, A., Sabater, J. M., Nicolas, J., Radu, R., Schepers, D., Simmons, A., Soci, C., & Dee, D. (2019). Global reanalysis: Goodbye ERA-Interim, hello ERA5. *ECMWF Newsletter* (159), 17–24. Available at: https://doi.org/10.21957/vf291hehd7

Islam, T., Rico-Ramirez, M., & Ishak, A. (2012). Performance evaluation of the TRMM precipitation estimation using ground-based radars from the GPM validation network. *Journal of Atmospheric and Solar-Terrestrial Physics, 77*, 194–208. Available at: https://doi.org/10.1016/j.jastp.2012.01.001

Jan, A., Baez-Villanueva, O. M., & Ribbe, L. (2022). *Spatio-temporal evaluation of gridded precipitation and evapotranspiration products over Balochistan Province in Pakistan.*

Kidd, C., & Huffman, G. (2011). Global precipitation measurement. *Meteorological Applications, 18*(3), 334–353. Available at: https://doi.org/10.1002/met.284

Salma, S., Rehman, S., & Shah, M. (2012). Rainfall trends in different climate zones of Pakistan. *Pakistan Journal of Meteorology, 9*(17), 37–47. Available at: http://www.pmd.gov.pk/rnd/rnd_files/vol8_issue17/4.pdf

Shahzad, N., & Amjad, M. (2022). Climate change and food security in Pakistan. In *Sustainable agriculture and food security* (pp. 579–594). Springer.

Tariq, A., Tabasam, N., Bakhsh, K., Ashfaq, M., & Hassan, S. (2014). Food security in the context of climate change in Pakistan. *Pakistan Journal of Commerce and Social Sciences (PJCSS), 8*(2), 540–550.

Ullah, S. (2017). Climate change impact on agriculture of Pakistan—A leading agent to food security. *International Journal of Environmental Sciences & Natural Resources, 6*(3), 76–79. Available at: https://doi.org/10.19080/IJESNR.2017.06.555690

Ullah, W., Wang, G., Lou, D., Ullah, S., Bhatti, A. S., Ullah, S., Karim, A., Hagan, D. F. T., & Ali, G. (2021). Large-scale atmospheric circulation patterns associated with extreme monsoon precipitation in Pakistan during 1981–2018. *Atmospheric Research, 253*, 105489.

Blue and Green: Hues of Riverine Flood Risk Reduction in Chennai, India

Jasjit Singh Banga, Yazid Ninsalam, and John Fien

1 Introduction

1.1 Grey to Blue and Green

Traditionally, flooding was considered to be the result of heavy rainfall or cyclonic events. Over a short period of time, this view has shifted through a recognition of the impacts of human actions. The hyper-urbanisation of mega- and especially coastal cities has triggered the transformation of environmental systems that were once seen as natural (Ahmed, 2018). Landfilling, urban developments, and building over rivers, floodplains, wetlands, and other waterbodies have disrupted the relationships of water and landscape (Dhiman et al., 2019; Manohar, 2016; Tajuddin & Dabrowski, 2021). Next to urbanisation, patterns of climate continue to change and exacerbate the severity of weather events (Gray & Ocampo 2017; Hamel & Tan, 2021; Manohar, 2016). These have weakened the capabilities of communities to cope with climate change and water-related hazards such as flooding. Addressing twin issues of climate change adaptation and disaster risk mitigation has become an 'urgent agenda' for urban professionals (World Bank, 2010).

Traditional ways of mitigating the impacts and resultant risks from urbanisation and climate change have involved human-engineered structures such as dams, piped drainage systems, concrete canals, seawalls, pumps, and water treatment plants commonly known as 'grey infrastructure'. Such tools may protect some communities from flood impacts. However, they cause major disruptions to the ecosystem and social functions and processes. For example, neighbouring communities and habitats

J. S. Banga (✉) · Y. Ninsalam · J. Fien
School of Architecture and Urban Design, Royal Melbourne Institute of Technology (RMIT University), Melbourne, VIC, Australia
e-mail: s3570895@student.rmit.edu.au

Y. Ninsalam
Biourbanism Lab, McGregor Coxall, Melbourne, VIC, Australia

N. Shahzad (ed.), *Water and Environment for Sustainability*,
https://doi.org/10.1007/978-3-031-27280-6_2

are often negatively impacted through the burning of fossil fuels, required to power the energy-intensive infrastructures such as wastewater plants and pumped drainage systems (Kabisch et al., 2016, p. 98). As a result, urban professionals are increasingly looking to environmentally adaptive or multi-functional approaches to flood risk mitigation (Gray & Ocampo 2017; Lafortezza et al., 2017). These include nature-inclusive approaches, such as ecosystem-based adaptation, ecological management, ecosystem and sustainable climate services, ecosystem-based disaster risk reduction (Kabisch et al., 2016; Lafortezza et al., 2017; Spyrou et al., 2021). Collectively, these are known as Nature-based Solutions (NbS) (Hamel & Tan, 2021; Lafortezza et al., 2017; Spyrou et al., 2021) as they have an ability to restore ecosystem health and overcome the limitations of hard engineering approaches to risk mitigation (Kabisch et al., 2016; Tajuddin & Dabrowski, 2021).

One of the key approaches to NbS involve Blue-Green Strategies (BGS): a sustainable, interconnected, and healthy network of water and landscape-based spaces and corridors (Hamel & Tan 2021; Kabischet al., 2016, p. 100). Blue strategies include the channels of water systems, such as wetlands, waterbodies, seas, rivers, and lakes in the built environment while green strategies involve the strategic use and placement of urban green spaces such as parks and gardens, as well as adjacent areas of paddies, mangroves, forests, and protected natural areas (Browder et al., 2019; Kabisch et al., 2016, p. 99; Klemm et al., 2017; Hamel & Tan, 2021). Combining blue (water) and green (landscape) provides multiple services and benefits to the ecosystem and the communities living in and adjacent to them, including the potential to increase resilience to water-related hazards.

1.2 Leapfrogging Towards BGS

One of the first approaches to BGS was the water-sensitive approach to city planning—Water Sensitive Cities (WSC). (Wong & Brown, 2009) articulated that such a planning practice can be achieved by capturing three principles: (i) potential sources of water emerging within the city, (ii) socio-ecological landscapes providing ecosystem services, and (iii) ecological and sustainable urban water governance. Water Sensitive Urban Design (WSUD) has advanced this notion by combining urban water planning with urban design to create blue and green networks that not only can support floodwater management but also assist in water filtration and biodiversity restoration, as well as enhance human health and food hygiene, and improve community living standards through activated public spaces (Wong & Brown, 2009).

Water Sensitive Urban Design (WSUD) and Blue-Green Strategies have been implemented and tested in a wide range of urban contexts in both the Global North and South. Projects such as RISE and Wetlands Work in Southeast Asia and the Pacific are notable successful examples in the Global South (Hamel & Tan, 2021). A wide range of demographic, climatic, economic, and governance issues have been found to influence the success or otherwise in these projects (Hamel & Tan, 2021). The most significant of these influences in South Asian and Southeast Asian cities is informal settlement. In the words of Satterthwaite et al. (2020) informal settlers often acquire land outside the formal jurisdictions of land ownership and urban services. Living without basic amenities and sanitation services which make them extremely vulnerable to climate change and hydrological disasters. This uncontrolled sprawl limits the design and implementation of NBS and even BGS. However, despite many studies showing that the integration of BGS is essential for flood mitigation, there exists no clear evidence on the practical performance of BGS in informal settlements (Hamel & Tan, 2021). This highlights the necessity of uncovering how BGS can be implemented in areas characterised by the complex interactions of informality with water and landscape.

This chapter responds to this need by assessing the implementation and performance of BGS in the informal settlements along the Adyar River in Chennai.

2 Land–Water-Human Relations in Chennai, India

2.1 The Context

Chennai (formerly known as Madras) is a coastal and capital city of Tamil Nadu, located on the Southeast coast of the Indian Peninsula. It is the fourth largest metropolitan area in India (populous) (Census of India, 2021; Greater Chennai Corporation, 2018). The Chennai Metropolitan Area covers 1189 square kilometres and includes parts of the Kancheepuram, Chengalpattu, and Tiruvallur districts, alongside Chennai city/district (Greater Chennai Corporation, 2018). Official plans to make Chennai the largest city in India by expanding its territories to 8800 square kilometres are underway (Vencatesan, 2021).

The Chennai Metropolitan Development Authority (CMDA) is the agency in charge of the Chennai Metropolitan Area (CMA) covering the districts of Chennai, Tiruvallur, Chengalpattu, and Kancheepuram. Chennai City/District covers about 426 square kilometres of urban area and is divided into three regions—North Chennai, Central Chennai, and South Chennai. It consists of 15 zones and 200 wards, which sits within the development and planning legislation of Greater Chennai Corporation (Greater Chennai Corporation, 2018; Saleem Khan et al., 2020).

Three major rivers, namely the Adyar, Cooum, and Kosasthalaiyar and the Buckingham Canal, flow through Chennai. The canal was created under the British colonial regime in the early nineteenth century and runs north–south through Chennai, parallel to the coast. This shallow canal connects the three rivers and was designed to provide drainage east to west in the city and promote flood mitigation (Biswas et al., 2019; Vencatesan, 2021). However, landscape and settlement changes in the two centuries since its construction have reduced its effectiveness as a flood mitigation measure.

There are three negative realities for Chennai: (a) a geographical problem (plain terrains) that makes the megacity invariably exposed to natural calamities, notably to floods; (b) the fact that despite being a coastal and wet city (1428 mm of annual rainfall), Chennai has grown with no regard to 'natural physiography' (Tajuddin & Dabrowski 2021) and without effective planning, especially regarding water systems and landscapes (Joerin et al. 2012a; Warrier, 2021); and (c) the axis of governance in Chennai and its parameters reflect tremendously over the landscape change and land-use pattern. The relationships and conflicts between the governing political party, State or Central, and the Corporation increase the tensions over implementing resourceful plans such as green and blue strategies (Tanner et al., 2009).

One of the geographical flaws that expose Chennai to natural calamities is that the city sits at an average elevation of 6 m from the mean sea level, which heightens its vulnerability to water-related hazards (Surendra, 2012). The natural water path has been disturbed by informal settlers, and approximately 150,000 people live along rivers, canals, and the coast (Esther & Devadas, 2016). They are the most vulnerable and at the forefront of riverine and coastal flooding.

To highlight the second reality, Chennai was initially planned with centralised man-made 'buffering mechanisms' (Dhiman et al., 2019), like lakes, pools, rivers, and temple tanks. These are known as the 'Eri-System', because 'Eri' is a man-made catchment to conserve rainwater (Tajuddin & Dabrowski, 2021; Vencatesan, 2021). These systems were constructed to restore water and mitigate the risk of flooding (Surendra, 2012; Vencatesan, 2021). However, with ongoing encroachments, hyper-urbanisation, and liquid and solid waste disposal, underlying water systems are losing their capacity to hold floodwater during medium to heavy rainfall (Biswas et al., 2019; Esther & Devadas, 2016; Warrier, 2021).

As such, the Eri system, which was created to protect the city from floods, has been suppressed by urbanisation and industrialisation (Dhiman et al., 2019; Joerin et al., 2012a, 2012b; Saleem Khan et al., 2020; Vencatesan, 2021; Tajuddin & Dabrowski, 2021). With the standout policy imposed by the Government of Tamil Nadu, around 90% of wetlands have been converted into residential, high-rise housing and urban developments, further destroying the city's blue and green network (Esther & Devadas, 2016; Vencatesan, 2021). In light of this, the term 'Poramboke' has been increasingly used among urbanists, engineers, and urban designers (Tajuddin & Dabrowski, 2021). The phrase in the Tamil language denotes open landscapes, which become wetlands or lakes upon rain, or a dry region known as no man's land—'Poramboke' (The Hindu, 2016). It was not until the 2015 floods—'Mega Floods'—that the government's attention was turned towards the consequences of the mismanagement of wetlands (Vencatesan, 2021). This makes it evident that such a terrible state of planning and wetlands mismanagement has negative social and economic impacts on the metropolis (Freestone, 2007; Vencatesan, 2021).

Finally, disagreements among fragmented governing bodies halt the implementation of Blue-Green Strategies that can mitigate disaster risks (Tanner et al., 2009). In most cases, new governments abandon previous government initiatives, killing off useful projects as a result (Tanner et al., 2009). This affects everything, including the delivery of essential services that communities need (Tanner et al., 2009). For instance, the housing scheme for Lower Income Groups (LIGs) was implemented by the Tamil Nadu Housing Board (TNHB). However, in the long term, the plan was unsuccessful, resulting in private developers purchasing numerous properties and converting them into multi-story buildings and eventually selling them with huge profits (Tanner et al., 2009). On the one hand, such disruptions and negligence by bureaucrats, who do not exhibit political inclinations, caused a waste of funds and, on the other, lessened the communities' coping capacities to extreme weather events. Consequently, these actions increase the city's physical fragility to risks and make it highly vulnerable to water-based extreme climatic events.

2.2 *Study Area*

This chapter targets three neighbourhoods of the Adyar riverbank but gives an overview of the scope, impact, and influences of urbanisation and related ecological changes in Chennai city.

With a keen interest in exploring various locations where the water-land nexus can be observed and documented, three locations were selected for investigation. A selection was made based on spatial quality, vulnerability, poor settlements, and neighbouring systems. The sites, namely (1) Maraimalai Adigalar Bridge/Saidapet Bridge, (2) Abraham Bridge, Saidapet, and (3) Burma Colony 3rd Street, Ekkatuthangal, Chennai, were selected (Fig. 1). While the latter—Burma Colony 3rd Street, Ekkatuthangal—was suggested by Social Activist, Nityanand Jayaraman, because of the fishing communities' vulnerability to riverine floods, the rest of the two locations were identified through basic geospatial analysis. Preliminary mapping revealed a picture of the complex edges of the land–water intersection. This prompted the authors to investigate the region and document the potential reality of landscape change in situ. As a result of these observations, the authors observed community behaviours that impacted and manipulated the relationship between water and land by using the Adyar river.

3 Methodology

This work was embedded in ethnographic research characteristics. Qualitative data collection methods included surveys, semi-structured interviews, and spatial, participant, and non-participant observational analysis. For instance, the pictures and diagrams narrated a story of the landscape changes that occurred, over the years, along the Adyar river. During the visit, Geographical Information System (GIS) informed the site selection criteria and presented statistical evidence of the changes. This research begins with stakeholder consultation and a survey, leading to a field

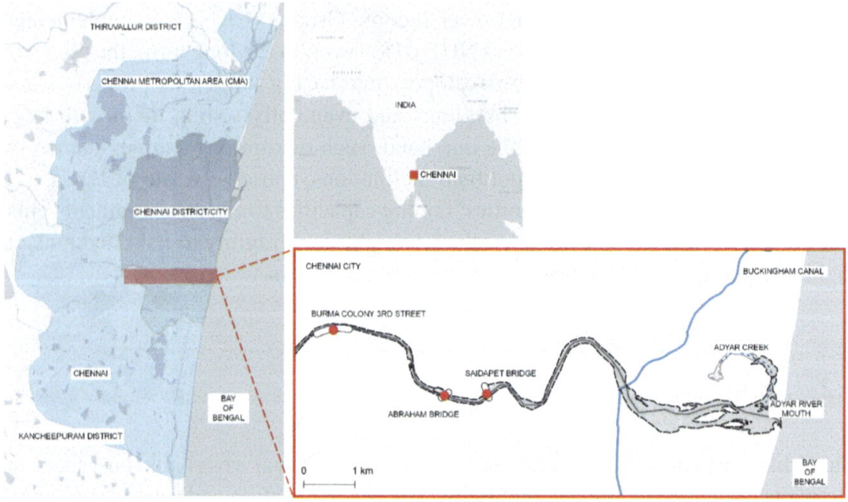

Fig. 1 Maps of study locations

visit to observe and document changes. A stakeholder is defined as an individual who may have a potential stake, direct or indirect, in a specific project. An informant, on the other hand, is the person or agency who informs or transfers the necessary information required to guide the study. This research acquired information from both sources.

3.1 Data Collection

The field visits conducted between 6 September 2021 and 12 September 2021 were aimed at gathering data on public–private space, river use, the relationship of local development with adjacent urban poor settlements, and waste management.

3.2 Stakeholder/Informant Survey

The authors have used vulnerability factors, defined by IPCC (2012)—social, environmental, spatial, and economic—to frame the questions of the survey. A preliminary stakeholder survey was circulated electronically among stakeholders and informants to identify the most vulnerable area—the hotspot area—where intervention is most needed. The semi-structured questionnaires were employed online to gather the necessary information regarding the neighbourhoods and communities residing around the flood plains of the Adyar river and the root cause of the riverine floods. This online tool was distributed among NGOs, academics and practitioners in the fields of Social Work, and Architecture and Urban Design who have a strong stake in flood mitigation projects in Chennai. However, we received five (5) responses (academics = 2, NGO = 1, practitioners = 2).

3.3 Semi-Structured Interviews

Informal interviews with the communities at Burma Colony 3rd Street allowed us to gain an overview of the residents living in the flood-prone neighbourhood of the Adyar river. While the other two sites, that is Saidapet Bridge and Abraham Bridge, Saidapet were explored and observed spatially, informal interviews at Burma Colony 3rd Street served as stops for the transect walk. At this stage, we aimed to understand the riverine floods from the perspective of the communities. The authors attempted to gain an empathetic understanding of the community's experiences through convivial conversations with residents. We used this lens to extract an overview of the strategies and policies imposed by the government upon them.

3.4 Transect Walks

At Burma Colony 3rd Street, the transect follows the main street in the neighbourhood, which runs parallel to the river and ends at a historically famous Hindu temple—'Aathu Kovil Muneeswaran' (Fig. 2). During this journey, we were given the opportunity to identify features and uncover information regarding diverse water-land intersections, the physical conditions of the river and its purpose. In so doing, we were able to gain the communities' perspective of the government's flood mitigation and public awareness strategies.

4 The Application of Blue and Green Strategies for Flood Mitigation

The survey allowed researchers to understand the aspects of government adaptation plans and policies that have been failing. The main reason for this failure is either the lack of rigorous study identifying the real-time problem or the implementation as per the communities (users) (Joerin et al., 2012a, 2012b; Saleem Khan et al., 2020). It is assumed by the authors that users are communities that live or work in the region and are primarily utilizing government-allotted services—directly or indirectly.

As one of the defining functions of disaster risk management, vulnerability may be positioned in the adaptation strategies of four sectors; spatial, social, economic,

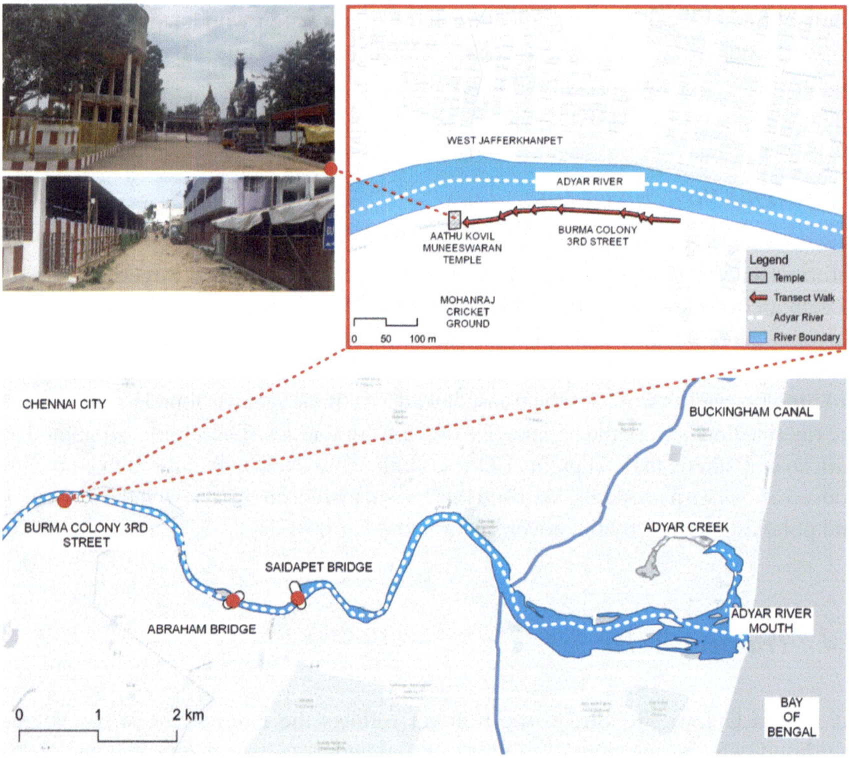

Fig. 2 Map of transect walks

and environmental (IPCC, 2012; Tajuddin & Dabrowski, 2021). The survey questions were categorised and linked accordingly. Additionally, an interviewee from the Burma Colony 3rd Street highlighted 'socio-political' as another factor in the context of Indian cities, especially in Chennai where waterbodies sit within 'fragmented governance' (Tajuddin & Dabrowski, 2021). Therefore, the barriers identified in this study are spatially, environmentally, economically, socially, and socio-politically bound and positioned.

4.1 Spatial Constraints

Highly increasing and dense urbanisation along with the terrible state of urban planning has impacted several spatial characteristics. Firstly, the city has grown with no effective planning and with no regard for waterscapes while neglecting the perspective of extreme natural calamities (Joerin et al., 2012a). The master plans, which favour fast-paced and commercially driven developments, cause a clash between natural ecology, hydrology, and geology. The rivers, canals, tanks, and lakes that were protecting the city from floods have been suppressed by urbanisation and industrialisation (Dhiman et al., 2019; Joerin et al., 2012a, 2012b; Saleem Khan et al., 2020). In addition to this, increasing demand for urban sprawl along the banks of the Adyar river has impacted the green network of the city. Because of this, communities are bound to live within tight spaces and afford minimal provisions for flora and fauna (Chaitanya, 2018).

The floodplains of the Adyar river are protected from the consequences of urban development. The banks are occupied by urbanity in the form of informal settlements, mostly situated on the riverbanks. In this context, Abraham Bridge represents a case of poor urban planning and improper land-use patterns (Fig. 3). We observe a massive waste collection centre located at the northerly end of the bridge which is located at the underpass and adjacent to the riverbank of the Adyar river (Fig. 4). Firstly, the geographical location of this waste hub poses a risk to nearby societies and systems that impacts the river and its natural ecology (Fig. 4). Similar to Saidapet Bridge, the edge of the water is delineated by the accumulation of solid waste, debris, and sand (Fig. 4e). The pile of waste cripples the ecology of the waterway in the form of a wall, temporary yet critical, that incapacitates the geo-hydrological function of the riverbank and makes it vulnerable to extreme weather events.

The study by Tajuddin and Dabrowski (2021) reveals that the walls, regardless of their spatial position and displacement of the original boundary, constructed

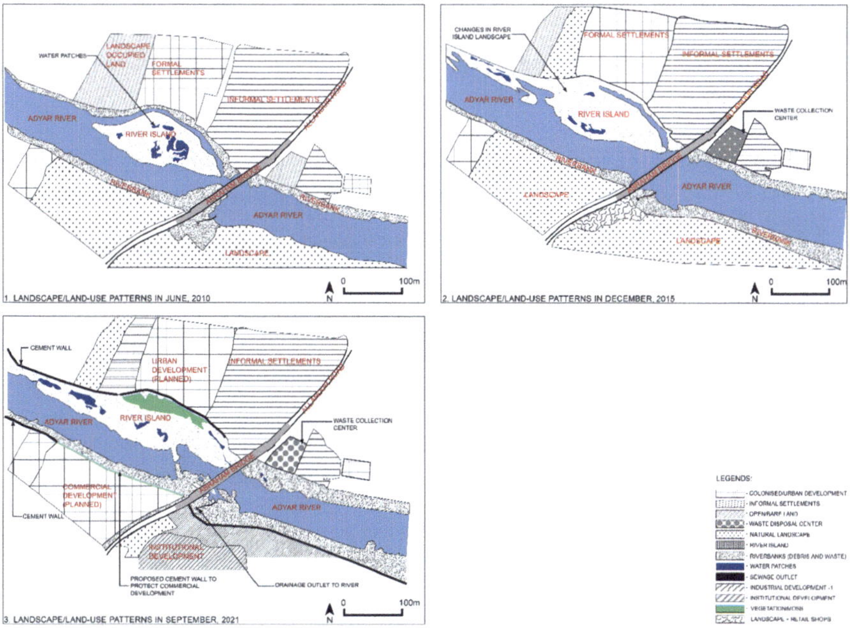

Fig. 3 (a) Map of the landscape patterns around the Abraham Bridge in June 2010. River water used to bypass the river islands, making it part of river ecology; (b) in December 2015, the landscape pattern of the river islands changed, blocking river water flow. *Source* Author

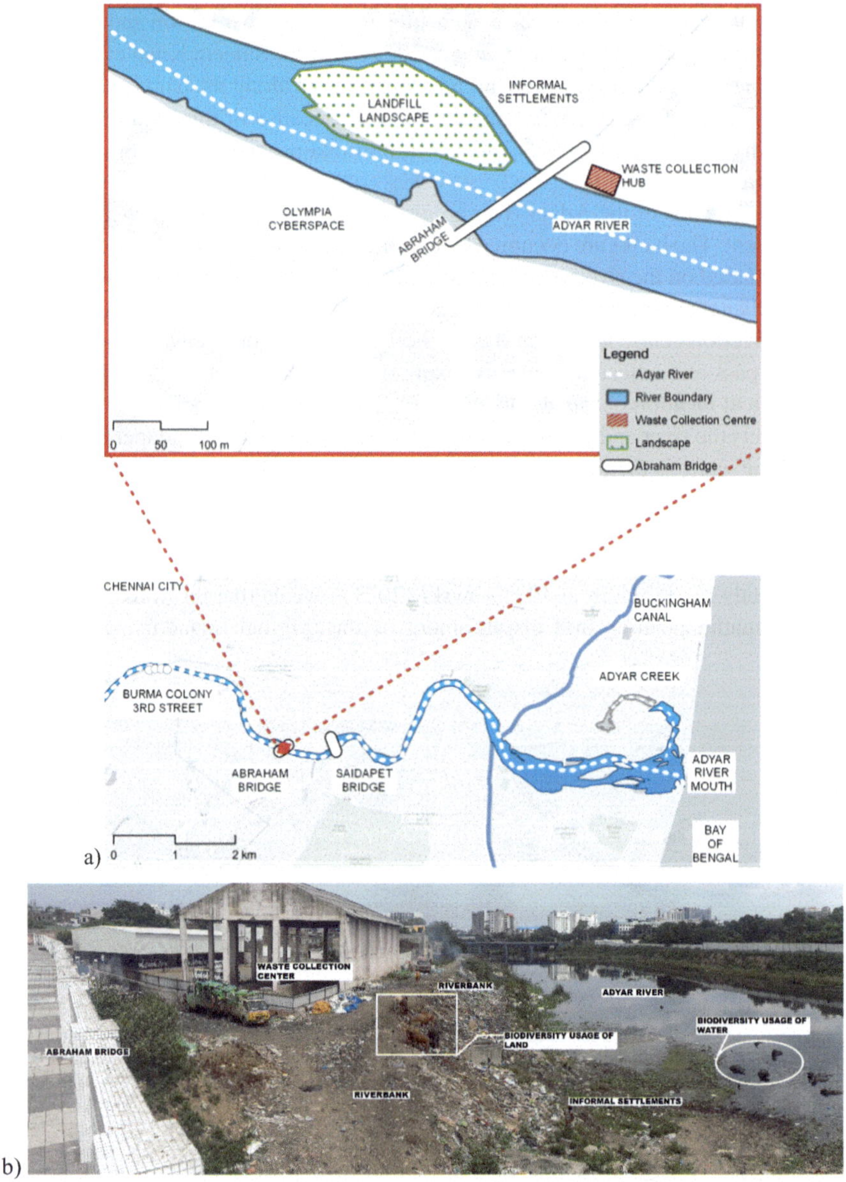

Fig. 4 (a) Map of Abraham Bridge, showing the state of poor urban planning which is destroying the river's flood-carrying capacity; (b) biodiversity using polluted river water for drinking and bathing and eating waste disposed of because of the spatial position of waste collection centre; (c–d) because of the waste disposal from informal and formal developments and urbanisation at the riverbanks, the landscape is interrupting the flow of the river water, resulting in frequent inundation; (e) untreated edges of water covered with temporary walls of sand, debris, and waste. *Source* Author

Fig. 4 (continued)

around the urban developments are labelled as the legitimate river boundary, whereas the urban poor settlements are considered encroachments. Because of this, informal settlers frequently face displacements which exacerbates their existing flood vulnerability (Tajuddin & Dabrowski, 2021).

The analysis of the 'Stakeholder Survey' and field observations revealed the spatial implications of the boarder study. While one of the stakeholders, Balaji Balakrishnan–Urban Designer and co-author of The Sponge Handbook-Chennai (2019), through a stakeholder survey, points out the 'fundamental problems, such as intervening at the regional scale', as the failure of the plans initiated for the improvement of the riverbank, others signify the understanding of landscape change at the neighbourhood scale before designing flood resilient strategies. The authors argue that ecological imbalances cannot be studied at a single scale and that interventions need to be made on a multi-scale basis.

Another example where urban development surpasses contextualised planning is the runway bridge at Chennai Airport. This runway, which is a part of urban development, is located at the pre-defined floodplains of the river and intersects the river at a low elevation level of 8 m above mean sea level (Fig. 5). This not only interrupts the flow of river water but also floods the runway with more than 10 m of flood level. With a flood level of 14 m, this region was inundated during the 2015 Chennai flood, while halting numerous domestic and international flights to and from the airport (Das, 2015). This brought a huge economic loss and restrained people from commuting within the country or internationally (Thiyagarajan, 2015).

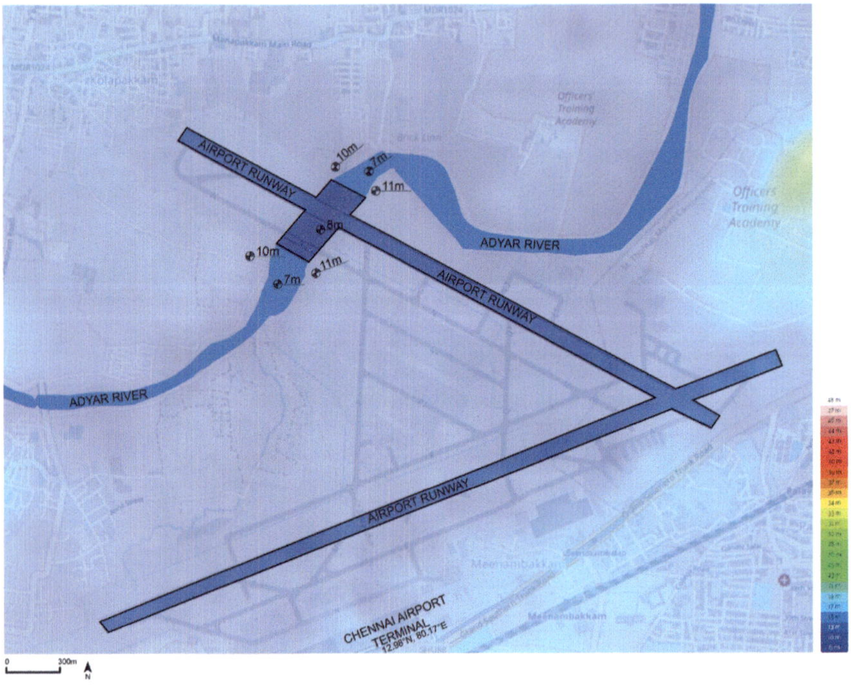

Fig. 5 Topographical map of Chennai's airport runway, which is constructed 1 m and 8 m above the river and mean sea level, respectively. It was inundated during the 2015 Chennai floods. Map adapted: Chennai topographic map

It illustrates the importance of urbanisation remaining within its boundaries and not occupying flood-prone areas. These damages are not only socio-spatially woven but also surrounded by economic and environmental destruction.

4.2 Environmental Constraints

Despite the differences listed above, environmental aspects such as climate change or poor environmental management expose communities and associated biodiversity to the risks of extreme weather events. The floodplains of the Adyar River are not protected from the consequences of urban development. Located along its banks are informal settlements that express urbanity.

There is a tremendous shift in the paradigm of river usage. Formal and informal developments interact with and use the river differently. In the former case, the river has become a sewage dump yard over time. It is definitely a mistake to make the river merely a spatial component that may or may not be pleasing to the eye. On the other hand, the latter was found to worsen the situation by dumping solid waste without anticipating the potential risks of riverine flooding. While the river serves as a medium of openness, open-air, and spatial access to water for urban development, settlements on the other side of the riverbank make up a picture of survival (Fig. 6d).

While conducting field visits, a water-land-human nexus was observed. It represents the mismanaged intersection between water and urban developments and rural or informal settlements. Submerged land and water edges are left untreated, and urban development has been protected by Grey Infrastructure—human-engineered structures. Invasive species grown in riparian zones may inhibit the flow of water into urban developments and reduce flood risks to some extent. However, on the other side, a temporary wall of sand, debris, and waste is erected naturally by dumping untreated solid waste. This structure is unable to withstand high or medium water flow. Apart from the landfill of debris and waste, the banks of the Adyar River hold the green belt of wildly grown species that are not adequately zoned and incapable of sustaining floods. However, these beds in conjunction with waste became a habitat for micro-organisms like algae insects, worms, and so on and a source of food for animals living in the area, such as cows, street dogs, and buffaloes (Fig. 4b). As a result of the accumulation of pollution from waste, the natural ecosystem is contaminated and the cycle of life is disrupted.

In 2015, in addition to heavy rainfall, Chennai was inundated because of excess water runoff into the Adyar river. As a result of this, the riverbanks were not able to contain the excess water leading to massive riverine floods in the surrounding regions (Shivakumar, 2018). The broken and shattered remnants of informal settlements that are in line with the riverbanks signal the consequences of past riverine floods (Fig. 6d). Residents prefer to look for temporary solutions to permanent problems instead of permanent aid. As a result, they seek short-term, flexible, and temporary 'Jugaad'. By adopting this, communities will seek temporary solutions as oblivion threatens their foreseeable future. Authors observed that unknowingly or perhaps knowingly, the communities (informal settlements) adjacent to the temporarily constructed and semi-structured infrastructure and residing or working on the floodplains have exposed themselves to the forecasted disaster of riverine floods.

The authors distinguished the poor condition of the existing waterway (Fig. 3b–e) and its interactions with humans and were surprised to witness landscape changes overlapping the water channel. The pockets of land comprising debris and solid waste forming in the river channel are blocking the hydraulic flow of water. This causes river water to overflow and flood its banks.

4.3 Economic Constraints

In Chennai, current urban paradigms revolve around the agenda of economic growth alone and neglect the uncertainties of climatic conditions. This neglect adds vulnerability to urban systems (Joerin et al., 2012a; Roggema, 2009). Discussion with stakeholders points to how government institutions prioritise the beautification of riverbanks solely to attract tourists and local migrants. This is at the expense of the

a)

b)

c)

Fig. 6 (a) Urban development and informal settlements at floodplains of Adyar River at Saidapet Bridge; (b) transformation of floodplain landscape to urban development which covers the river with its construction and waste at Abraham Bridge, (c) sewage outlet of formal settlements to Adyar river, (d) signs of survival from previous floods and disposing of waste by informal settlements at Abraham Bridge site. *Source* Author

d)

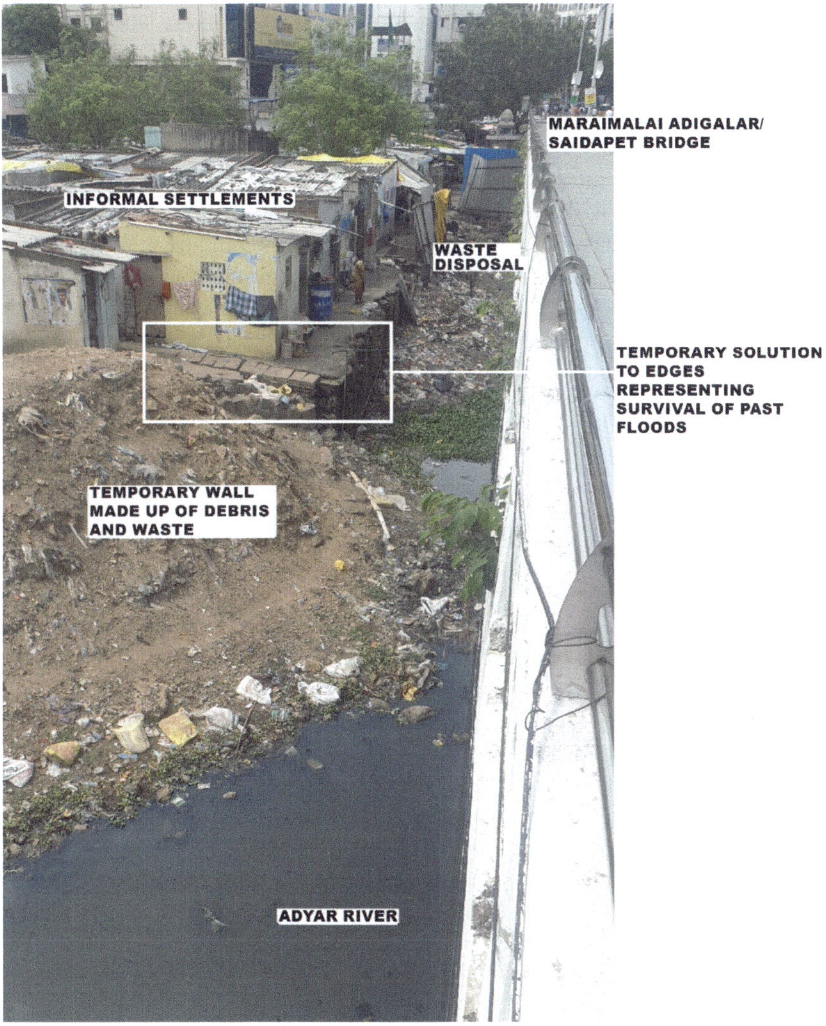

Fig. 6 (continued)

Adyar River's natural ecology and biodiversity. Field observations, a transect walk, and the survey have revealed that the riverbanks of the Adyar River are occupied mainly by informal settlements on the one end, and urban developments, like five-star hotels, apartment buildings and complexes, located on the opposite riverbank.

Urban developments are considered as an indicator of economic development whereas informal settlements are seen as illegal encroaching activities. It illustrates local authorities' preferences for development typologies along riverbanks. Despite this, a large number of urban poor live in flood-prone slum settlements around the Adyar river (Joerin et al., 2012a). The informal settlements comprised of fishing and boating communities living in encroachments over flood plains are labelled as 'urban poor' (Gray & Ocampo, 2017; Joerin et al., 2012b). These 'urban poor' contribute massively to the water-based trade to and from the coast of Chennai.

The master plan developed and implemented by the Chennai Metropolitan Development Authority (CMDA) shows signs of fast-paced economic benefits through the development of urban infrastructure around the river edges (Tajuddin & Dabrowski, 2021). However, this is detrimental to the hydrological regime located within the catchment. The scheme attracts private investment by highlighting the spatial benefits while avoiding the impacts it may have on neighbouring societies and ecology. Moreover, the plan lacks a strong background analysis, for example, the mapping and documentation of informal entities in the first place and the ecosystem services rendered by waterbodies. In their study, Tajuddin and Dabrowski (2021) emphasize the importance of 'market-driven' and 'water protection' urban planning at the regional level.

4.4 Social Constraints

Three locations were chosen for the fieldwork in order to depict the differential and tangled relationships between humans and water and land. Urban development dominates one end of the river, while informal settlements occupy the other end, forming an urban poor township. Based on the data gathered from the responses (number = 5) of the online survey, many of the fishing communities situated around the neighbourhoods of the Adyar River are the most vulnerable to riverine floods. In contrast in adjacent societies—urban poor, airport infrastructure and Middle-Income Groups (MIG)—people are equally susceptible to floods.

Though the nominated regions are distant from each other, surprisingly, informal settlements, urbanisation, and severe rainfall events are the root causes of floods. Flood vulnerability is exacerbated by the lack of adaptation strategies implemented by the government and people.

Dependency on others, especially on government agencies for survival, could be linked to low 'participation at the local level' (Tajuddin & Dabrowski, 2021). The government authorities hold a significant position in Indian cities in controlling the various systems that emerge. As a result, the vulnerable community of Chennai depends on government agencies and other stakeholders not only for their resilience and adaptation measures but also for their needs (Joerin et al., 2012b). Indirectly or directly, this dependency increases communities' challenges and exposes them to social conflicts.

On the ground, the urban poor's social differences often create a barrier to resilience. Their minimal participation in adaptation activities and limited knowledge about landscape change, including waterbodies, contribute to this. There is a need for awareness in regard to the 'value of urban water' and crippling ecosystems (Tajuddin & Dabrowski, 2021).

As one of the interviewees at Burma Colony 3rd Street revealed, there is a huge gap in knowledge sharing. This is because of insufficient involvement or amendments by government agencies relating to such communities. As such, this neglect can make these settlements even more vulnerable to the impacts of climate change and other hazards (Joerin et al., 2012a). When the interviewee outlines adaptation strategies related to floods, it appears that there is a correlation between livelihood and resilience levels. The residents of the underlined neighbourhood have lived there for more than 20 years and have witnessed the growth of urbanisation and the changes that have occurred. River water's performance impacts their livelihood and well-being and that of the next generation. As described by one of the interviewees, during the 2015 Chennai floods, the water entered their homes and reached the upper floor level. Yet, there are hardly any major amendments or implementations of flood mitigation strategies in this neighbourhood. This provides an excellent example of how government policies are solely the manifestation of a political and economic agenda. These adaptation strategies often fail to bridge the social differences between urban and rural, formal and informal, and rich and poor communities (Gray & Ocampo, 2017; Joerin et al., 2012b). We observed that the lack of awareness of government or organisational initiatives and strategies, like Adyar Eco-Park, Resilient Chennai, Sponge Collaborative, Cities of 1,000 tanks, and so on among these communities, amidst the compounding negative landscape changes arising out of climate change and hyper-urbanisation (Fig. 3), could be a major factor for the vulnerability of particular social groups and their exposure to natural hazards. Additionally, in Chennai, while preparing or developing such strategies, government leaders tend to neglect these communities in their design and decision-making process (Joerin et al., 2012b; Tanner et al., 2009). Therefore, these pieces of evidence suggest that existing flood mitigation strategies have been ineffective.

4.5 Socio-Political Constraints

The socio-political deterioration begins with the lack of knowledge possessed by the communities and the mismanagement of the water bodies (Tajuddin & Dabrowski, 2021). Chennai is dealing with the ownership barriers. The governing hierarchy and authorship of the water bodies is unclear. In order to regain the resiliency of the region, there needs to be an immediate answer to who owns the water bodies, and bare lands, and who manages them (Tajuddin & Dabrowski, 2021). Fragmented governance operates waterbodies, canals, and rivers. Different departments operate and manage water bodies, canals, and rivers, which is confusing for locals to understand and approach. This creates a long-term barrier to sociocultural resilience to floods in the city.

The word 'Poramboke', which means 'no man's land' (The Gopalakrishnan, 2017; *The Hindu*, 2016), has recently become increasingly popular among water engineers, urban planners, and designers about the waterbodies in Chennai. After a rain event, a wetland that is filled withby rainwater becomes a waterbody, whereas when it is dry, it is known as a 'Poramboke' (*The Hindu*, 2016). Since the English East India Company and the British rule, from an economic perspective, these lands have been crowned as 'wastelands' since they do not generate revenue (Gopalakrishnan, 2017). In the stakeholder interview, Chennai River Restoration Trust highlighted the inappropriate labelling of these lands and stressed the importance of these buffers for flood mitigation (Tajuddin & Dabrowski, 2021). However, because of mismanagement, rivers and water bodies have been encroached on by urban developments and exposed the region to flooding.

Fieldwork reveals that communities living over riverbanks are either unable to understand the ecological imbalances occurring or government agencies have failed to educate them. One of the interviewees at Burma Colony expressed the failure of government initiatives as the strategies are insufficient or irrelevant to the real-time context. Regulations that protect 'sensitive ecosystems from rapid urbanization' have undermined the river's resilience.

Observations of private, public, and government ownership were also examined. The floodplains of the Adyar river are managed by different institutes. For instance, CMDA holds accountability for master planning and post-flood relief works, whereas Chennai Rivers Restoration Trust is formed by the Tamil Nadu (state) government to develop an 'Eco Park' model for the restoration of riverbanks in Chennai. Both entities work individually and within their parameters instead of working together. That is why a lack of knowledge sharing has been observed within the communities as well as within the responsibilities of the governing agencies.

The development of new programmes like Smart City (Aijaz, 2021), in most cases, diverts the focus of the governing bodies from on-ground issues. A stakeholder's informal interview revealed that the World Bank and the State Government agencies have proposed such programmes, but the workshops were held in a five-star hotel and were inaccessible and uncomfortable to members of vulnerable communities. This made it difficult for the communities to attend these events and present their needs to stakeholders. The agencies aimed to restore the aesthetics of the riverbanks while neglecting the ecological balances and the perspectives of urban poor and vulnerable communities. As such, this represents a lack of initiatives aimed at working together with communities and understanding their real-time problems and needs.

5 Recommendations for Programmatic and Context-Based Landscape Designs

The key contributions of this study relate to how the principles of Nature-based Solutions, specifically the characteristics of Blue-Green Infrastructure, can be linked with the landscape shift in the neighbourhoods of the Adyar River. We have observed and learnt from the communities' actions and their engagement with the local landscape at Saidapet Bridge, Abraham Bridge, and Burma Colony 3rd Street. We use these learnings to contextualise the practical applications of blue-green systems and produce recommendations for communities to bridge into programmatic and context-based landscape designs:

5.1 Developing with Nature

The survey highlighted the value of Nature-based Solutions (NbS) and the name of institutions that might be carrying forward its practice. The project, namely the City of 1000 Tanks, 'identifies the interrelationships between the causes of floods, water scarcity, and pollution that focuses on providing a holistic solution to the underlying problems in Chennai' (Ooze, 2019). As a result of hyper-urbanisation, the Mughals suppressed and invalidated the water circulation strategy of the seventeenth century. This system considered how nature could guide urban development and could inform the principles of Water Sensitive Urban Design (WSUD) through a NbS approach.

Chitra Nagar in Kotturpuram, which was nominated by stakeholders in the survey, describes a project that decentralizes the use of NbS to save, collect, and recharge water catchments. Considering water as a medium, the design idea reforms the existing informal fabric into a typical high-rise housing building while anticipating 'rainwater harvesting, small waste treatment plants, uniform open spaces, solar power, and gender-sensitive toilet design' (Prabhakar, 2021). The practice focuses on catching and retaining water to use during the dry season for drinking (Prabhakar, 2021). In other words, the strategy aims to solve the challenges of water supply as Chennai is at risk of losing its groundwater because of urban development covering the city's water-storing mechanisms. However, the proposal is mismatched with the context of local landscapes and green infrastructure specific to Chennai city. Specifically, it focuses on water retention and supply issues and avoids sociocultural barriers to flood mitigation. The realities of this project no doubt attract private investors but at the expense of the displacement of communities and biodiversity along the river, while prioritizing the goals of river beautification.

With the backdrop and blend of blue and green systems, we recommend revising and restructuring the existing fluvial flood adaptation strategies. So far, various local and international institutions have been established by the government to develop new plans for river restoration. Sadly, this agenda neglects the existence of informal settlements and the perspectives of its inhabitants with the river. We recommend that future initiatives should run parallel to the needs of local low-income communities. The lens of landscape-based approaches could make the urban poor pivotal in planning, waste recycling, and restoring river ecology, enabling people and nature to live in harmony. For instance, these Poramboke or 'wastelands' (Fig. 7) can be reclaimed by installing bioretention pits, fish farming systems, bio-filter planting, and swales. This plan would restrain the urban poor from dumping the waste into such pockets and refresh the ecology of the river.

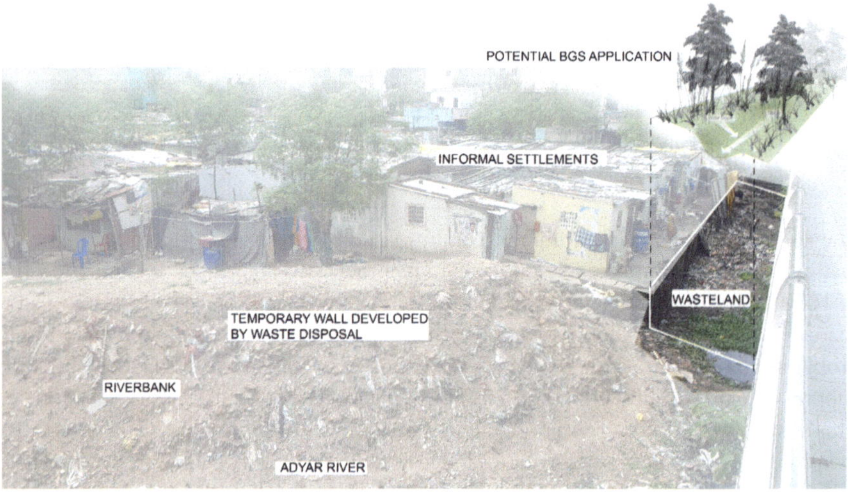

Fig. 7 BGS can retain and restore water and mitigate the risks of floods on riverbank's wastelands

5.2 Living with Nature

Based on on-site visits (three locations), we understand that the communities want to make nature a part of their life and live with it, but on the ground, in terms of their living culture, such implications have failed. This might be because of the scarcity of resources or perhaps because of the lack of knowledge sharing. It has been accounted for that the informal or urban poor communities living along the Adyar River are unable to understand the purpose of the river. As well as dumping waste, they use the river to wash clothes, do laundry, and for sanitation purposes. As a consequence, the physical environment is degraded, the river is polluted, and the flow of the river is obstructed. Because of this, during the yearly monsoon, the water of the river overflows, resulting in flooding the riverbanks. Future projects should leverage the implementation of BGS by developing alternate approaches and perspectives towards the existence of the river. Through natural elements, wastewater from the river could be treated and filtered and recirculated. In this way, NbS projects would be able to restore the river's ecology, improve urban ecosystems and protect biodiversity.

5.3 Designing with Nature

A lack of access to essential services, especially during emergencies, threatens the resilience of communities. By synthesising spatial data acquired during the field visit, we documented various small alleys and dead ends. These alleys which were not captured by digital satellite mapping systems, such as Google maps and others. Due to the lack of appropriate and accurate spatial information, urban planners and designers have not been able to include existing fabric to inform their design proposals. Ultimately, this leaves the communities vulnerable to low-to-high-impact development projects.

In light of this, manual identification of narrow streets, alleys, and more could assist in analysing the complexity of existing fabric at the neighbourhood level. For example, a spatial understanding of the site informed by (McHarg & Lewis Mumford, 1969) and Rem Koolhaas and Bernard Tschumi's entry for the Parc de la Villette competition that utilises mapping of landscape systems could guide planners, designers, and engineers to generate layers of functions that consist of blue-green and social systems. Ultimately, design resolutions could be derived from these ecological and cultural layers. The dead spaces and alleys can be utilised to restore the ecology while providing amenities for the people living in the region. As such, the sustainably developed patches in the neighbourhoods would form a network of blue-green infrastructure, representing resilience at the regional scale.

5.4 Limitations Afforded by Nature

From field observations, we observed that the infrastructure of informal settlers grew organically. Building lines are unplanned and impenetrable, resulting in narrow lanes. Blocks of shacks are built within tight spaces without following accessibility guidelines. In order to accommodate maximum utility placement within spatially constrained spaces, the houses are constructed vertically—with three to four floors covered by tarpaulin sheets or metal roofs. The surfaces upon which their houses are built are shattered and cracked that reveal signs of past flooding events. We doubt that these structures can withstand further fluvial or perhaps pluvial floods in the future. The tangled characteristics of informal settlements could outperform the performance of various human-engineered green infrastructures. Due to complex and structurally inadequate housing structures, for example, green roofs (engineering systems) are invalid and inapplicable. To prevent local pluvial floods, alternatives such as naturally planted community gardens with water retention systems installed at abandoned spaces (Fig. 7) can be used. With BGS, the climate adaptation of informal settlements could be upgraded with the development of open and cool spaces and increased water services (Satterthwaite et al. 2020).

5.5 Communicating the Importance of Nature

Communities living along the riverbanks of the Adyar are mostly involved in fishing activities as a source of income to sustain their livelihoods. However, these communities are unfamiliar with the impacts of urban development that may impact ongoing riverine activities. The reason for this is a lack of awareness and formal education to inform them of its consequences. For the government, it is a crucial task to spread such awareness among the urban poor. This is to build the capacity for communities in understanding social vulnerability to both anthropogenic and climatic events. We recommend localised governance where a community leader can be assigned to guide and educate a small to medium number of informal settlers. The dissemination of information through movies and visual aids is an influential medium to spread awareness. Such mechanisms would leverage the work led by social activists, community leaders, and non-governmental organisations to convey and educate the practical benefits of BGS to the residents of these informal settlements. This is especially relevant for fishing communities, which depend on the ecosystem services provided by blue-green systems as their livelihood relies on nature.

5.6 Stewardship of Natural Resources

In this study, the in-situ gap is identified as the absence of community participation in the design process. This could be a key factor contributing to the ongoing failure of existing plans and strategies. By observing people's behaviours and actions towards the 'river's ebb and flow' (Prescott et al., 2022), we can understand their perspectives and day-to-day interactions with the river. In this stage of research, we perceive the river to be a convenient place for the urban poor to dump their waste. Poor spatial and infrastructural conditions, and challenging sanitation conditions that release untreated wastewater, further exacerbate residents' vulnerabilities to climate change and extreme weather events. Despite such ecological and social challenges, communities remain isolated, compounded by the lack of practical flood resilience and mitigation strategies. In order to fill this gap, future research should propose that vulnerable communities play an active role in planning and designing resilient strategies. Community-centred and co-design approaches benefit society in two ways. As a result, stakeholders are able to identify communities' (users') problems in real time, and as a result, people can invest themselves and accept the responsibility for their actions. Synthesising such data helps to build comprehensive bounce-back strategies and community masterplans.

Informal interviews with the fishing community revealed that, although the Tamil language was the native language of Chennai, previous workshops and seminars were conducted in the English language. Moreover, these were organised in inaccessible locations for the communities to participate meaningfully. As a solution, we suggest that scholars must first acknowledge and understand the traditions and cultures of the city and its communities. Community-centred workshops, seminars, and conferences must be organised with a grounded approach to making them accessible to the local communities. In case of cross-language constraints, a local guide and translator should be appointed to communicate the ideas and discussions clearly during such programmes. These user-friendly events will not only make people feel comfortable but will also motivate them and attract neighbouring communities to actively participate and achieve mutual goals.

6 Conclusion

This chapter highlights various constraints that make the Global South countries incapable of implementing disaster risk reduction strategies. Spatial, environmental, economic, social, and socio-political constraints are major factors in effective flood risk reduction.

As well as a lack of effective city planning, urbanization has exploded like mushrooms due to spatial constraints. Cities in the Global South are adversely affected by hyper-urbanization. As the pace of urbanisation increases rapidly in a coastal metropolis like Chennai, heavier, frequent, and devastating floods become a norm and impact the social and environmental systems of the city. In light of this, context-based planning enables us to study and intervene at multiple scales in the context of complex ecological imbalances.

As noticed in Chennai, the listed environmental constraints highlight the inefficiency and futility of grey Infrastructure. Adoption of such an expensive system may protect urban developments and assist in managing urban water catchments, but it impacts adjacent societies and biodiversity. In other words, grey infrastructure disrupts the ecosystem and may create environmental limitations to flood mitigation in the long run. On the contrary, nature-inclusive approaches could upgrade the adaptation of communities and the habitats through the effective implementation of BGS.

Similar to the cities in the Global South, the agenda of the urbanising model in Chennai is to achieve fast-paced economic gains. Rivers and the surrounding environment are at risk due to the focus on attracting private investors and maximising investments. It is common for plans to disregard the existence of informal settlers living in the riverside neighbourhoods and focus on aesthetic appeal. Considering this, it would be crucial to develop maps containing many layers that gather information on various ecological and social characteristics. This system would develop context-oriented or site-specific design resolutions of Nature-based Solutions (NbS) to address the complexities of urban expansions in the Global South.

Added on to this, as a social constraint, it has been noticed that the community was not considered or paid attention to while developing the solutions. Without community consultation and a baseline understanding of the larger context of the landscape have often been inadequate as an effective barrier against flooding. From observations and recommendations, a clear picture emerged, showing that nature-based solutions (NbS) should complement engineering solutions for flood risk reduction in Chennai, particularly through the consideration of Blue-Green Strategies (BGS).

This chapter has identified fragmented governance and ownership as the crucial part of socio-political limitations to flood mitigation. Not only does the danger of flooding grow with the re-engineering of natural landscapes, but it is also made more severe by governance issues and a lack of proper mitigation awareness among urban settlers (specifically informal, poorer settlements). On a governmental level, clarity must be achieved about the ownership of related blue-green assets. On a community level, effective communication of management strategies and practical ways of communicating strategies across communities living along the rivers, regardless of demographic strata, is key. An inclusive approach that combines government cooperation, community awareness, and Blue-Green Strategies (BGS) would be a viable next step in the Global South for reducing the impacts of potential flood events.

Acknowledgements We would like to thank the respondents to the survey for their time and input in this research. The findings presented herein are based on the views of these respondents and authors. We would also like to acknowledge the contributions of Prof Lawrence Surendra and Dr D. Narasimhan for their inputs on data collection and reporting and guidance. We are grateful to Dr Bridget Keane, TSP Asia, and RMIT University, Australia for supporting this research and providing academic resources.

Disclosure Statement

No potential conflict of interest was reported by the author(s).

Additional information

Funding This research was supported by the Australian Government through the Australian Alumni Grant Scheme (AAGS) 2021. We would like to express our great thanks to Australian Government and the Australian Consulate-General Chennai for providing us financial support.

Appendix

Stakeholder survey

1. Nominate and share the link of a neighbourhood using https://what3words.com/fortheinterventionaroundAdyarriver,accordingtoitsvulnerabilitytoriverine floods.

2. Typeofcommunitiesresidingorworkingintheregion? *Mark only one oval.*

   - Urban poor
   - Elderly
   - Fishing
   - Agricultural
   - Other:

3. Causesoffloodsinthenominatedneighbourhood? *Check all that apply.*

   - Urbanisation
   - Informal settlements
   - Poordrainagesystem
   - Overwaterreleasefromdams
   - Grey infrastructures (human engineering)
   - Severe climatic events
   - Waste disposal
   - Other:

4. Whatincreasesvulnerabilityofthecommunitiestofloods? *Checkallthatapply.*

   - Social conflicts Dependency on agencies
   - Natural calamities
   - Hyper-urbanisation at flood plains
   - Imbalanced social strata
   - Lack of adaptation strategies
   - Physical disability
   - Other:

5. Towhatextent,povertyandsocialinequalitycontributetowardsvulnerabilityof the communities to riverine floods? *Mark only one oval.*

   - Low
   - Moderate
   - High
   - Severe

6. To what extent, climate change and poor environmental management contribute towards vulnerability to the riverine flooding? *Mark only one oval.*

- Low
- Moderate
- High
- Severe

7. To what extent, vulnerable informal livelihoods and their dependency on governmentagenciesorstakeholderscontributetowardscommunities' vulnerability? *Mark only one oval.*

- Low
- Moderate
- High
- Severe

8. To what extent, ground surface elevation or contours and unregulated land use planning contribute towards vulnerability to riverine flooding? *Mark only one oval.*

- Low
- Moderate
- High
- Severe

9. Areyouawareofanyplansforimprovementtotheriver?*Mark only one oval.*

- Yes
- No

10. Doestheplanisrelevantandadequate?Pleasegivereasons.

____

References

Ahmed, F. (2018). Flood vulnerability assessment using geospatial techniques: Chennai, India. *Indian Journal of Science and Technology, 11*. https://doi.org/10.17485/ijst/2018/v11i6/110831

Aijaz, R. (2021, August). *India's smart cities mission, 2015–2021: A stocktaking* (ORF Special Report No. 155). Observer Research Foundation.

Biswas, S. K., Raj, P. R. S. L., & Balaganesan, B. K. P. S. (2019). *The Sponge handbook: Chennai—Using the landscape approach to transform the South Buckingham Canal Area, GIZ.*

Browder, G., Ozment, S., Rehberger Bescos, I., Gartner, T., & Lange, G.-M. (2019). *Integrating green and gray: Creating next generation infrastructure*. World Resources Institute. http://hdl.handle.net/10986/31430

Census of India. (2021). *List of mega cities In India 2021|Top 5 populous city in India.*

Chaitanya, S. K. (2018). Chennai's green cover shrinking, needs structural plan, finds report. *The New Indian Express*, viewed 20 December 2021.

Chennai Rivers Restoration Trust (CRRT) 'Adyar Park'.
Das, M. (2015). Warnings not to build runway over Adyar river ignored, that's why Chennai airport sunk. *The News Minute.*
Dhiman, R., VishnuRadhan, R., Eldho, T. I., & Inamdar, A. (2019). Flood risk and adaptation in Indian coastal cities: Recent scenarios. *Applied Water Science, 9*(1), 1–16. https://doi.org/10.1007/s13201-018-0881-9
Esther, S., & Devadas, M. D. (2016). A calamity of a severe nature: Case study—Chennai, India. *Urban Water III, 1*(Friar), 227–236. https://doi.org/10.2495/UW160201
Freestone, R. (2007). The internationalization of the city beautiful. *International Planning Studies, 12*, 21–34. https://doi.org/10.1080/13563470701346527
Gopalakrishnan, S. (2017). *No man's land.* India Water Portal. www.indiawaterportal.org, viewed 26 December 2021.
Gray, S., & Ocampo, M. A. (2017). Resilient edges: Exploring a socio-ecological urban design approach in metro Manila. *The Plan Journal, 2*(2), 519–561. https://doi.org/10.15274/tpj.2017.02.02.17
Greater Chennai Corporation. (2018). *CDMP Book Wrapper full book (English).pdf.*
Hamel, P., & Tan, L. (2021). *Blue–green infrastructure for flood and water quality management in Southeast Asia: Evidence and knowledge gaps.* Environmental Management. https://doi.org/10.1007/s00267-021-01467-w
IPCC. (2012). *Special report on managing the risks of extreme events and disasters to advance climate change adaptation (SREX).* http://preventionweb.net/
Joerin, J., Shaw, R., Takeuchi, Y., & Krishnamurthy, R. (2012a, May). *Action-oriented resilience assessment of communities in Chennai, India* (Vol. 7891). https://doi.org/10.1080/17477891.2012.689248
Joerin, J., Shaw, R., Takeuchi, Y., & Krishnamurthy, R. (2012b). Assessing community resilience to climate-related disasters in Chennai, India. *International Journal of Disaster Risk Reduction, 1*(1), 44–54. https://doi.org/10.1016/j.ijdrr.2012.05.006
Kabisch, N., Frantzeskaki, N., Pauleit, S., Naumann, S., Davis, M., Artmann, M., Haase, D., Knapp, S., Korn, H., Stadler, J., Zaunberger, K., & Bonn, A.(2016). Nature-based solutions to climate change mitigation and adaptation in urban areas and their rural surroundings. *Ecology and Society, 21*(2). https://doi.org/10.5751/ES-08373-210239
Klemm, W., Lenzholzer, S., & Van Den Brink, A. (2017). Developing green infrastructure design guidelines for urban climate adaptation. *Journal of Landscape Architecture, 12*(3), 60–71. https://doi.org/10.1080/18626033.2017.1425320
Lafortezza, R., Chen, J. Q., van den Bosch, K., & Randrup, T. B. (2017). *Nature-based solutions for resilient landscapes and cities.* Special section on Nature-Based Solutions (NBS) for resilient landscapes and cities: Innovative strategies and scientific advances (Vol. 159). https://doi.org/10.1016/j.envres.2017.11.038
Manohar, L. (2016). Towards resilience in Chennai. *Towards Resilience in Chennai, 17*(3), 251–260. https://doi.org/10.7480/iphs.2016.3.1266
McHarg, I. L., & Mumford, L. (1969). *Design with nature.* American Museum of Natural History.
Ooze. (2019). *City of 1,000 Tanks, Chennai*, viewed 29 December 2021.
Prabhakar, S. (2021). Chennai: Here's how the urban poor can live in a dream home—Chennai News. *Times of India*, viewed 29 December 2021.
Prescott, M., Ninsalam, Y., & Luo, R. (2022). Learning from Kampungs in Jakarta: Contextualising water sensitive cities principles with community tactics and behaviours. *Landscape Architecture Journal, 29*(2), 26–38. https://doi.org/10.14085/j.fjyl.2022.02.0026.13
Resilient Chennai 'Resilient Chennai Strategy Kaleidoscope'.
RISE 'Revitalising Informal Settlements and their Environments (RISE)'.
Roggema, R. (2009). *Adaptation to climate change: A spatial challenge.* Springer.
Saleem Khan, A., Sabuj Kumar, M., Sudhir Chella, R., & Devdyuti, B. (2020). *Chennai city and coastal hazards: Addressing community-based adaptation through the lens of climate change and sea-level rise (CBACCS)* (pp. 777–798). https://doi.org/10.1007/978-3-030-37425-9_39

Satterthwaite, D., Archer, D., Colenbrander, S., Dodman, D., Hardoy, J., Mitlin, D., & Patel, S. (2020). Building resilience to climate change in informal settlements. *One Earth, 2*(2), 143–156. https://doi.org/10.1016/j.oneear.2020.02.002

Shivakumar, C. (2018). Video: Why Chennai floods—The News Minute. *newindianexpress*, viewed 3 January 2022.

Spyrou, C., Loupis, M., Charizopoulos, N., Apostolidou, I., Mentzafou, A., Varlas, G., Papadopoulos, A., Dimitriou, E., Panga, D., Gkeka, L., Bowyer, P., Pfeifer, S., Debele, S. E., & Kumar, P.(2021). Evaluating nature-based solution for flood reduction in spercheios river basin under current and future climate conditions. *Sustainability (Switzerland), 13*(7), 1–20. https://doi.org/10.3390/su13073885

Surendra, L. (2012). Chennai, India. In *China, India, and East and Southeast Asia: Assessing sustainability* (pp. 42–45). Berkshire.

Tajuddin, N., & Dabrowski, M. (2021). Enabling socio-ecological resilience in the global south: Insights from Chennai, India. *Sustainability (Switzerland), 13*(19), 1–24. https://doi.org/10.3390/su131910522

Tanner, T., Mitchell, T., Polack, E., & Guenther, B. (2009). *Urban ggovernance for adaptation: Assessing climate change resilience in ten Asian cities* (IDS Working Papers, No. 315, pp. 01–47). https://doi.org/10.1111/j.2040-0209.2009.00315_2.x

The Hindu. (2016). "Poramboke" song plays back reasons for flooding—Chennai. *The Hindu-Chennai*, viewed 26 December 2021.

Thiyagarajan, S. (2015). *Chennai airport is doomed to suffer, its runway is built on river.*

Vencatesan, J. (2021, September 1). *Hang in there, Chennai!* Urbanet.

Warrier, S. G. (2021). *Chennai floods again after losing many blue and green spaces [Interview].*

WetlandsWork. *Flood prone villages—Wetlands work!.*

Wong, T. H. F., & Brown, R. R. (2009). The water sensitive city: Principles for practice. *Water Science and Technology, 60*(3), 673–682. https://doi.org/10.2166/wst.2009.436

World Bank. (2010). *Cities' contribution to climate change* (pp. 14–32). Urban Development Series.

Water Balance Equation for Rivers of Assam, India

Pranjal Kumar Phukan

1 Outline

Rivers contain a significant amount of water, which is regarded as a natural resource that is essential to the sustenance of life and the environment. The majority of the population in India is dependent on rivers, which play an important role in the country's economy. Rivers serve as a significant hub of biodiversity. However, it has been noticed recently that river water is now polluted in several ways. When toxins from groundwater affect water, it is referred to as being contaminated. Both people and aquatic life are severely suffering as a result of these toxins. Since surface and groundwater sources are constantly degraded by municipal dumping, residential waste, and industrial waste, people are becoming more concerned about the quality of these resources as the population grows. Additionally, the percolation of this waste or dirty water is contaminating the groundwater aquifer and also damages the soil, which results in the pollutants lowering the soil's quality. Other significant sources of surface and groundwater pollution include runoff from agriculture, chemical leaks, sewage leaks, and dangerous microorganisms.

Located in the north-eastern region of India, the state of Assam has a total land area of 30285.00 square miles and is bounded by latitudes of 24°50′ to 28°00′ north and longitudes of 89°42′ to 96°00′ east. Assam experiences a relatively warm and humid monsoon season. The Brahmaputra River traverses all of Assam in an east to west direction.

In their 2017 study on the worsening of the Bharalu River's water quality, Narayani Gogoi and Rashmi observed that whenever a pollutant enters the river, a variety of physical and chemical, and biological events are activated, and the consequence is mostly influenced by the type and amount of the pollutant. The analysis revealed that due to the extensive dumping of organic materials in bed, the river has been

P. K. Phukan (✉)
CosmosMinds, Dibrugarh, India
e-mail: pranjal_xlri@yahoo.com

N. Shahzad (ed.), *Water and Environment for Sustainability*,
https://doi.org/10.1007/978-3-031-27280-6_3

transformed into a stagnant water body with high levels of free carbon dioxide toxicity and hydrogen sulphide (H_2S).

The effects of the pollution of some significant rivers in India and the corrective actions taken were reviewed by A. K. Panigrahi and Subasini Pattanaik (2019). Based on the research, the main source of pollution in India's rivers is industrial effluents and metals, including Fe, Hg, Cd, Pb, and Zn. The findings of studies on several river water quality measures were reported.

The water quality of the Bharalu River in Guwahati was thoroughly examined by Mahananda Borah (2020). The examination of the river's water pollution used a variety of methodologies. The preliminary study, field research, experiments, and interpretation and analysis were the four stages of the study. The types of pollutants and the variations in contamination levels were investigated.

The study of the hydrology of a basin or region is best approached through the water balance once the measurements of the hydrological cycle's constituent parts are complete. Simply put, this stipulates that the amount of water input into a region must balance its output and any changes in storage over any given time period. The primary source of input is probably rainfall, though there is a chance that rivers or groundwater would also enter the system; the primary sources of output include evaporation, river outflow, and groundwater outflow; and the primary sources of storage changes likely increase in soil moisture and groundwater storage. Balance sheets are one type of example of a water balance study, but other examples include conceptual models that convey a physical understanding of the connections between various processes.

Water balance is the bedrock of management and policymaking in any crucial matter involving water resource management. Depending on the goal and level of precision, its models can be on various time scales, such as hourly, daily, monthly, or yearly.

The water balance equation, in its ideal form, reads as

$$\text{Input} = \text{Output} \pm \text{Change in Storage}$$

Overflow starts when there is insufficient storage or outflow. The magnitude of the spillover increases with the net positive variance between inflow and outflow. It would be ideal to undertake action to minimise runoff and improve storage for the long-term sustainability of a city like Guwahati in Assam, India, which relies on pumping stations. Fortunately, several steps have already been taken, and continuing efforts are underway. For the purpose of reducing runoff, authorities may also take into account enforcing permeable paving and rainwater harvesting for residents of the city.

2 Methodology

When utilising remote sensing, there are several ways to estimate the water stress in a specific area. The most crucial parameter to evaluate water stress, according to Jackson et al. (1981), is surface temperature. The presumption behind the relationship between water stress and surface temperature is that as plants transpire, the evaporated water decreases the temperature of the leaves below that of the surrounding air. Thus according to Moran et al. (1994), although the water deficit index, or WDI, is essentially the ratio of actual to prospective evaporation, it can be feasibly calculated using remotely sensed surface temperature, reflectance (red and near-infrared), along with limited onsite meteorological data.

3 Water Balance Analysis

The Bharalu River flows through the Indian state of Assam and is a tributary of the Brahmaputra River. Before joining the Brahmaputra River, the Bharalu River passes through the heart of Guwahati, where it derives its source in the Meghalaya Khasi Hills. The Bharalu River represents one of Assam's most polluted rivers.

There is insufficient information on the biodegradability, toxicity, or carcinogenicity of any of these substances, although there is no doubt that health concerns have risen over time. Urban runoff has had an influence on a river's water quality as it traverses through an urban area. Due to crowded locations, urban runoff is extremely polluted and contributes significantly more contaminants to stream flow than river base flow or sewage effluents.

The North-Eastern region's rivers, which are dominated by the Brahmaputra and its numerous tributaries, have not been well studied both for water chemistry and pollution. Communities and decision-makers in this region generally agree that the water of these watercourses (rivers) of the River Brahmaputra has maintained its undisturbed purity to this day since this region (North-Eastern States of India) is well known among financial experts and captains of industry in the country for its industrial backwardness. Since many of the state's rivers are still regarded as sacred by the inhabitants, they have become an essential component of religious rites, cultural legacy, and social dynamics in Assam. Second, environmental awareness in this region of the country is relatively new, and very few studies on various aspects of water pollution in Guwahati have been published. As a result, the discharge of untreated sewage and other municipal trash has been neglected throughout the process.

The Bharalu River (Fig. 1) is Guwahati's natural drainage system. It collects sewage and other waste from local businesses, restaurants, hotels, schools, and other public spaces before flowing to the Brahmaputra at Bharalumukh. The National River Conservation Directorate (NRCD) has set a permissible limit of 3 mg/L while the river's biochemical oxygen requirement is 52 mg/L. Furthermore, Guwahati inhabitants are at risk for health problems due to the river's offensive odour.

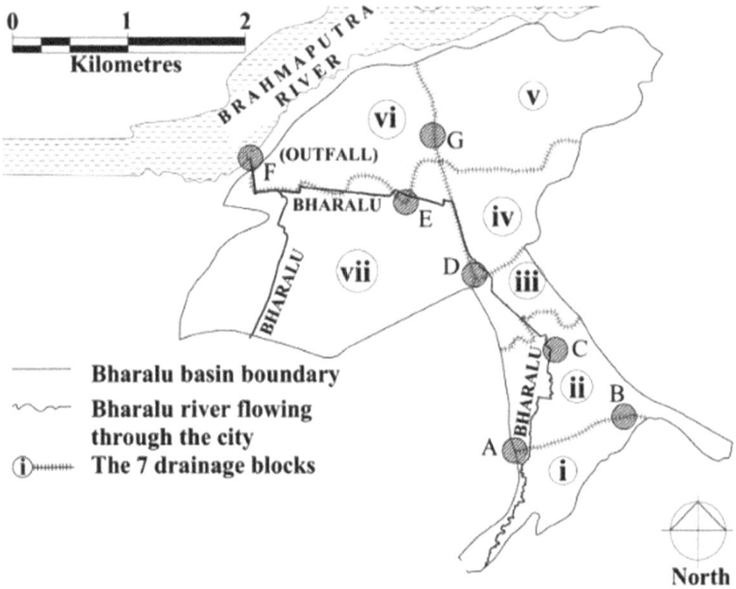

Fig. 1 Bharalu river

For the Bharalu, once the sluice gate is closed, this water balance equation can be written as:

$$\text{runoff} + \text{waste water} = \text{discharge}$$

where discharge is from the pumping station [with $\pm$ change in storage].

This equation can be simplified as:

$$\text{runoff} = \text{discharge[from the pumping stations]}.$$

This simplified equation demonstrates that when the carrying capacity is sufficient, artificial flooding in the city can be primarily caused by an imbalance between runoff and outflow. Nevertheless, the carrying capacity may be deemed adequate as long as there is no discernible change in the level differential between the sump and any upstream portion during the pumping station's maximum capacity operation.

The Bharalu has an 11-square-kilometre catchment area, according to a piece of public domain literature. With a runoff factor of 0.8, 5.5 mm of precipitation causes this catchment area to lose 48,400 cubic metres of water each hour. Additionally, the installed capacity of the Bharalumukh pumping station is limited to this amount of runoff or roughly 5.5 mm of rainfall.

The Bharalu water balance can be reviewed in light of the aforementioned information to target a relatively accurate yield. The findings of such an investigation can aid in understanding the current situation and the creation of action plans to control the variables with the ultimate objective of controlling the spill.

Untreated sewage from the city gets transported through the Bahini-Bharalu River and discharged into the Brahmaputra River. Unhygienic sewage also reaches an important bird sanctuary. According to industry analysts, the city now generates roughly 70 million litres of sewage per day (MLD), but that number will inevitably rise as new water supply systems are installed in the Guwahati Metropolitan Area. By 2025, it is projected that the amount of sewage produced would reach approximately 280 MLD.

With a BOD level of 52 mg/l and Guwahati's sewage as the primary cause of pollution, the CPCB has categorised the Bharalu River as one of the "priority 1" categories of polluted river segments. The National Green Tribunal (NGT) directed the development of action plans to "restore the polluted river stretches to the prescribed standards" in September 2018 after taking note of the 45 critically polluted stretches across the nation (as classified by the CPCB), which included the Bharalu River in Guwahati. The action plans must, among other things, address the control of residential and industrial pollution at the source, the channelisation and disposal of treated domestic sewage, the management of river basins, the regular inspection of groundwater quality, etc.

Reviewing the water balance will also reveal whether the pumping station is enough. The pumping station may be thought of as the city's lifeline for maintaining a flood-free environment due to the terrain of the city and the surrounding area. Here are some comments on the pumping station in the light of those considerations.

3.1 River Bharalu's Leading Pollutants

1. Bharalu River pollution is mostly caused by biochemical oxygen demand (BOD), which is one of the major pollutants. BOD is produced by a variety of sources, including leaves, woody debris, dead animals and plants, animal manure, effluents from pulp and paper mills, wastewater treatment facilities, malfunctioning septic systems, and urban stormwater runoff.
   - Desirable limit: 4 mg/l
   - The limit in Bharalu River: 12 mg/l Limit
2. Fluoride: The emission of industrial effluents containing inorganic fluorine compounds causes fluoride pollution in the Bharalu River. In addition to being utilised as a flux in the steel and glass fibre industries, fluorine is also used in the manufacturing of aluminium. They are also released when phosphate is manufactured.
   - Desirable limit: 1 mg/l
   - Maximum permissible limit: 1.5 mg/l
   - Limit in Bharalu River: 0.02 mg/l to 3.73 mg/l

3. pH: Variations in river water's pH are brought on by activities like agricultural runoff, acidic mine drainage, including emissions from fossil fuels like carbon dioxide, which, when dissolved in the water, produces a weak acid. The pH of water has to have a considerable impact on the aquatic flora and fauna as well as on water quality. Aquatic organisms will succumb if the pH of the water is either high or too low. The solubility and toxicity of chemicals and heavy metals in water can also be affected by pH.
   - Desirable limit: 7–8.5
   - Maximum permissible limit: 6.5–9.2
   - pH limit in Bharalu River: 6.8–7.7
4. Dissolved oxygen (DO): Water-based plant and animal life have always needed dissolved oxygen to survive. Reduced DO is a sign of the existence of what is known as oxygen-demanding wastes, which are chemicals under pressure. Low DO values are a sign that there is a lot of biodegradable organic matter present, which makes the water look unsightly because the anaerobic breakdown of organic matter results in septic conditions and the creation of unpleasant fumes. By regulating reduction/oxidation reactions and the leaching of metal ions from the soil, DO also affects the equilibrium between water and sediments.
   - Desirable limit: 6 mg/l
   - Maximum permissible limit: 4 mg/l
   - DO limit in Bharalu river: 0.5 mg/l – 1.25 mg/l

3.2 Corrective Actions to Reduce Pollution

The restoration of the river environment depends on the strategy that is chosen. Because of the high levels of pollution, cleaning up polluted rivers is a pressing problem in many emerging economies. Pollution control at the source point or in-situ water treatment is two different ways to remediate polluted river water.

3.3 Methods to Lessen Bharalu River Pollution

A. To collect the sewage from both sides of the river, interceptor pipes shall be installed along the Bharalu.
B. No industry should discharge its effluent straight into drains without treatment; alternatively, it should reuse its treated effluent or sewage.
C. Identifying communities for sewage treatment plans and building sewer systems.
D. The disposal of untreated sewage and solid waste into surrounding drains and rivers by local households or hotels/restaurants should be prohibited.

E. Public education campaigns to prevent open defecation and increase awareness of hygienic practices.

Mitigation actions to lower pollution in the river encompass:

Several corrective actions must be conducted to lessen the effects of the pollutants/compounds:

1. Biochemical Oxygen Demand (BOD): By incorporating hydrogen peroxide into the wastewater solution, BOD can be lessened. The observed BOD will be reduced as a result of the hydrogen peroxide's chemical attack on the organics in the wastewater.
2. Fluoride: Water is exceedingly difficult to purge of the chemical component fluoride. To lower the fluoride content in water, we can utilise electrodialysis and dialysis.
3. pH: The pH of wastewater gets decreased using appropriate acidic chemicals. The most inexpensive and easily available is sulphuric acid. It is greasy, thick, and severely corrosive. Wastewater containing acid has indeed been mitigated utilising sodium hydroxide and calcium hydroxide.
4. Dissolved Oxygen (DO): The primary cause of low dissolved oxygen is the phosphorus-induced overgrowth of algae. The growth of algae is also influenced by nitrogen, another nutrient. Dissolved oxygen is used in the algae's decomposition and death process. By boosting wave and wind movement, putting plants in the water, and exposing that to pure oxygen, one could boost the amount of dissolved oxygen in the water.

3.4 Reserve Pumps

This is the only mechanism, for obvious reasons, that, once the sluice gate is closed, allows excess water produced by continuous rain to be discharged. If placed properly, a sufficient number of backup pumps with independent setups identical to the existing ones might be extremely beneficial. If the old pumps need to be replaced with a bigger, more energy-efficient pump with a stronger suction lift, a cost–benefit analysis can be used to evaluate whether it's feasible. It is also possible to look at the statutes of the priming system. Even if it is determined that the capacity of the current pumps is enough after assessment, they can be taken into consideration. In addition, additional decentralised pumping stations can be considered to take care of the flood-prone areas.

3.5 Continuity of Operations

In order to keep the sump free of unwanted debris and sustain the functioning and/or efficiency of the pump at all times, suitable screens can be installed, preferably along the last section of the river. If every one of us who resides in the catchment area is on alert, the frequency of cleaning these screens can be reduced considerably. In order to ensure that all pumps are available for continuous operation, the power supply and distribution system, including drivers and switches, must be maintained and improved as warranted.

Affected parts of the pumping system could be those exposed to corrosive wastewater. This issue can be minimised by applying appropriate paints on a routine basis.

To ensure that it delivers continuous service throughout the monsoon, every such appropriate precaution should ideally be made during the lean season.

A pump's operation may not always demonstrate that it is functioning properly. Tracking the parameters that indicate the output of the pirating pumps permits the timely correction of any problems. Since the cause of the overflow could be ascertained if the relevant and important parameters could be maintained in real time both before and during pump operation, a review of the facilities available for monitoring and recording the critical parameters of the pumping station, including the level of the Brahmaputra River, may be carried out.

3.6 Pump/Sluice Gate Operation

By adjusting how well the pump station and sluice gate operate, experiments can be carried out. It is important to identify and publicise the sump danger level at which flooding in the catchment areas occurs. The sump's level, including the rate at which its water level is rising, should be carefully watched during periods of severe rain. To maintain the water level inside the danger zone, both the pumps and the sluice gate should be closed beforehand. Such a mind-set will also handle the rise in the Brahmaputra's water level.

3.7 Debris Handling

The solids handling process is not only arduous, expensive, and time-consuming, but it also irritates the wider populace. Any sources of debris penetration into the channel can indeed be located and closely monitored. For the purpose of raising awareness and encouraging compliance, the public can be informed about the contribution that city people can play in completely or partially eradicating such intrusion. To maintain the entire waterway clear of debris, especially during the monsoon, ongoing efforts are required. In order to minimise any negative impact on the functioning and effectiveness of the pump, emphasis should also be made on maintaining the pump and sump clean of any undesired debris.

Artificial flooding has the potential to significantly disrupt urban life. In addition to the loss of property and maybe lives, there is a significant loss of man-hours that affects the daily activities of all segments of the population. If indirect losses are also taken into consideration, the amount lost may be significant. In such circumstances, residents not only face a risk to their health and hygiene but are also denied access to vital public services.

According to experts, climate change makes it more difficult to predict heavy rainfall, which reduces our ability to prepare for catastrophic weather events like floods in the future.

In order to assist authorities in tackling this difficult issue during the monsoon, a professional organisation might be hired to monitor important parameters in key Bharalu-Bahini system locations using contemporary tools and methodologies.

As per the Indian Constitution, it is everyone's responsibility to conserve the environment, including rivers, forests, and other natural resources. To get our city back to being flood-free, we should endeavour to see that this clause is adhered to.

4 Monitoring Approach

A substantial portion of the city's effluent has customarily been transported mostly through the Bharalu. Except for some slum dwellers, the Bharalu water has long ago discontinued being used for drinking. However, the water has been utilised for other purposes, 'like washing clothes, watering lawns, washing outside premises of factory/industry establishments etc'. The Guwahati community has long expressed concern over the Bharalu water's worsening qualities through local publications and other venues.

The chosen approach should be to periodically monitor the parameters so that an overall picture of the water quality of the Bharalu River could be obtained, and based on this, future research of an extensive nature could be motivated.

5 Conclusion

Based on the water parameters obtained at several hydrological stations along the Brahmaputra River, the study's overall objective was to evaluate the evolution of water quality at spatial and temporal scales. According to the pollution sources listed, anthropogenic factors such as population growth, excessive consumption, overuse, chemical explosions, and subsurface mining are the main causes of pollution in the catchment area. One of the main causes of this is a general lack of awareness among the general public. Another factor is the absence of government policies and initiatives to protect the river ecosystem. As a result, appropriate measures must be made to restore the river ecology, which, if not preserved, will result in a major ecological imbalance. While significant species of flora and wildlife along the banks have already vanished as a result of high pollution levels in a relatively short period, adequate steps must be taken to protect the ecosystem that is still present. Therefore, the tributaries' inputs of contaminants to the river must not exceed the necessary levels. The Bharalu River, a tributary of the Brahmaputra, as well as its connections with the beels (wetlands) and the Brahmaputra River, make up Guwahati's natural drainage system. The Mora Bharalu, Bahini, and Basistha are only a few of the numerous tiny rivers that feed into it. Currently, the Bharalu collects a sizable percentage of the city's municipal, industrial, and other wastes and acts as the natural drainage for stormwater runoff. The study concludes that pollution in the Bharalu River can be significantly decreased if the relevant steps stated above are carried out suitably. Rivers are natural resources, so it is important to take the necessary steps to prevent pollution.

Numerous examples like these show the general idea that if the climate, physical hydrology, and vegetation are carefully taken into account, simple water balance models can be very effective at explaining the physical processes taking place in a river basin and can lead to an understanding of runoff. A fundamental understanding of different hydrological regimes can be gained from comparisons of precipitation and potential transpiration, but seasonal timing is just as significant as annual totals. Simple water balance models may be incredibly helpful in explaining the hydrology of a region and can act as a helpful check on more complex hydrological modelling and fundamental data when combined with knowledge of the key driving forces. It is crucial to make sure that the hydrological forces trying to control runoff are understood and incorporated into the analysis in these days of increasing assessment complexity and reliance on computer data manipulation, rather than relying on more complex sets of parameters and relationships to explain runoff.

References

Panigrahi, A. K., & Pattanaik, S. (2019, July). A review on consequences of pollution of some Indian major rivers and their remedial measures. *International Journal of Research and Review, 6*(7). ISSN: 2349-9788.

Pollution Control Board, Assam, Conservation of River Bharalu, Preparation of detailed project report, 2013.

Effects of Structural Measures for Flood Mitigation on Migratory Behaviour of the River Indus Using Geographic Information System and Remote Sensing: A Case Study

Fatima Afzal, Naeem Shahzad, and Syeda Uzma Naureen

1 Introduction

Numerous geomorphological and river management issues require a thorough grasp of how river channels have changed throughout time. The term "lateral migration" describes how a river channel shifts position in response to changes in fluid flow and sediment discharges. This process is invariably accompanied by bank erosion of the stream bed or channel wall when there is turbulent flow. Therefore, lateral migration is a process that has the potential to result in drastic local or regional changes (Yang et al., 1999). Instead of using integrative strategies like expanding river space and halting construction on flood-prone terrain, traditional flood risk management largely relies on structural methods to reduce flood danger (Galloway & Lewis, 2012). In reality, because the public believes that flood risk has been reduced, structural interventions like levees (dikes), dams and constructed channels often encourage urban and agricultural expansion on the floodplain (Serra-Llobet et al., 2022). The livelihood of people is hampered by intermittent minor flooding, which also makes the environment less stable. In order to create stability and predictability, structural measures like adding levees and dams have been a common method of flood management. This traditional wisdom, however, has been called into question by the finding that removing regular flooding is frequently linked to an area's greater susceptibility to infrequent floods over time (Sung et al., 2018).

River channels evolve naturally with the passage of time. Such a change is most prominent in the lateral section where it is observed in either size or shape of the river. Lateral shift in the large sedimentary rivers has gained massive attention in the fields of hydrology, geomorphology, geology and engineering (Dunne & Leopold, 1978).

F. Afzal · N. Shahzad (✉)
National University of Sciences and Technology, Islamabad, Pakistan
e-mail: naeemshahzad@mce.nust.edu.pk

S. U. Naureen
CMH, Risalpur, Pakistan

N. Shahzad (ed.), *Water and Environment for Sustainability*,
https://doi.org/10.1007/978-3-031-27280-6_4

The work started from simple defining the formation of channels of a single alluvial river (Leopold & Wolman, 1957) to complex phenomena of meandering of the rivers and its modelling to predict the effects of features and form in the fluvial system (Lancaster & Bras, 2002). Many factors are responsible for this cross-sectional shift for example underlying geology, hydrologic management, land use and land cover, etc. Soil erosion is topping the list for land degradation and river channel migration in Mediterranean region. This process is directly responsible for flooding in the region and shifting of the river segments. Further this process is aggravated by development of hydrologic catchment variables in the fluvial environment for flood management (Poesen & Hooke, 1997).

Images from Landsat (MSS and TM), SPOT, RADARSAT, NOAA, IRS and many more lightweight satellite systems are useful sources of satellite data for river investigations. For fluvial applications, higher resolution data are preferred. Landsat data are the only multi-spectral digital data with synoptic coverage dating back to 1972, despite the fact that its spatial resolution is not as precise as that of SPOT HRV data (Yang et al., 1999). Geographical Information System and Remote Sensing are a useful technique in quantification of erosion rates of the Brahmaputra–Jamuna River. It used a time-based analysis of the remote sensing data from 1973 and 1992 and measured the lateral shift of river at regular intervals (Khan & Islam, 2003). Land-use change detection using satellite remote sensing and geographic information systems (GIS) has proven to be a strong and reliable process. Over the past few decades, it has been extensively used to monitor coastal diversity and river deltas (Chen et al., 2005).

River Indus is one of the most important rivers of south Asia due to its irrigational system and economic importance in Pakistan. River Indus having a complete length in excess of 2800 km originates from Tibetan Plateau in close vicinity of Lake Mansar over beyond Himalayas and flows through three highly agricultural provinces of Pakistan, namely Khyber Pakhtunkhwa, Punjab and Sindh and finally falls into Arabian Sea in the south. River flows smoothly through the Indus Plains. However, the natural course of the river has been altered by anthropogenic modifications. During the inundation season, to ensure proper water supply off-season and to reduce the devastation by floods, flood water has been stored in many river segments and levees that do not allow the natural evolution of river path and have forced the river to change its natural course. One such site is a river segment at Ghauspur Daho which is situated ~61 km of river length south of Guddu barrage and is ~32 km long with a prominent mid-channel bar formation. By using Landsat images from 1972 to 2013, we have studied the general trend of migratory behaviour of river and MODIS imagery, with high temporal resolution, for year 2003, 2005 and 2010 have been employed to observe the flood water residence and its impact on inland migration of the river channel. The analysis demonstrates the effects of disaster mitigation practices on the Indus water channel near Ghauspur Daho.

Considering the anthropogenic drivers in enhancing the erosion rate of the river banks, the river no more evolves naturally, rather is pushed away from baseline making the floodplain unstable and vulnerable to further hazards. Unsustainable structural measures taken to mitigate flood inundation risk have increased the water residence time of flood water increasing the bank erosion and consequently shifting the river laterally. Evidential study based on facts and figures is required to advocate the need of sustainable structural flood mitigation measure. Main objective of the research is to study the lateral shift of Indus River at Ghauspur Daho and establishing a relation between the flood water residence time, as result of structural measures to mitigate flood and the enhanced rate of erosion of river bank.

2 Study Area

The focus area selected for this study was the Ghauspur Daho region. This is situated ~61 river km downstream of the Guddu barrage at the Punjab-Sindh border and extends down south in a southwesterly direction for about ~93 river km on the western flank while on the eastern flank the river length is ~90 km. It is highly pertinent to mention that during the *floods of 2010*, it was at stream segment to the west that a major breach happened. This breach resulted in widespread destruction in the western districts of northern Sindh extending to the piedmont of the Suleiman-Kirthar ranges (Fig. 1). Water further drained down towards the Manchar Lake along the piedmont causing additional damages.

Additionally, on the western fringes of the selected segments are a series of small reservoirs meant for basic mitigation and off-season water supply storage pools (Fig. 2). These evidently effect the erosion and stream evolution on the western

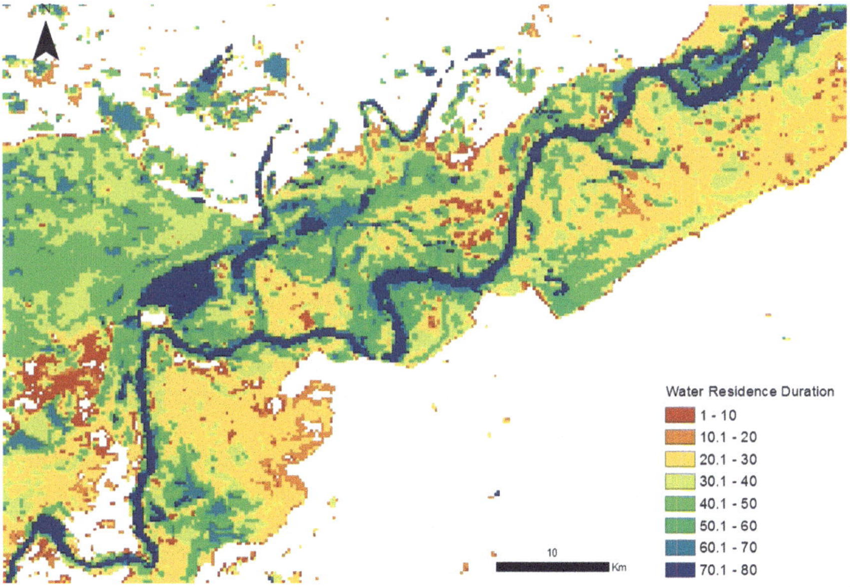

Fig. 1 Breaching in the northern parts of the Sindh province along the western Indus River bank

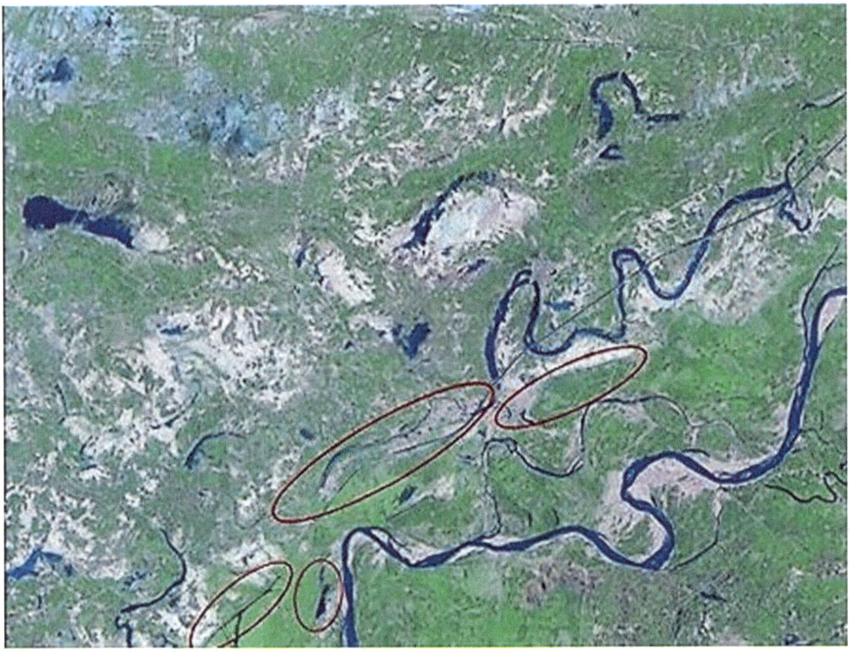

Fig. 2 Storage pools on the western shore of the Indus River near Ghauspur Daho

fringes. Another mitigation feature near the stream segment is ~90 km long Kandhkot old stream channel (Fig. 2) which is also used for excess discharge and off-season irrigation uses.

3 Remote Sensing Analysis

Landsat and MODIS satellite remote sensing data were used in analysis of flowing and stagnant water due to their high to moderate resolution and easy availability. Previous studies in hydro-geomorphology reveal that their resolution both spatial (Landsat-30m and MODIS-250m) and temporal (Landsat-16days and MODIS-32days) are the best suited to such studies. Landsat images are good for stream characteristics' mapping, while MODIS images are good for high temporal resolution flood residence time mapping.

One of our objective was to map the river stream followed by the inundation duration near the Ghauspur. For mapping the river channel, Landsat images were considered suitable in the light of existing studies (a study of the disaster mitigation practices and its effects on the migratory behaviour of the Indus River channel near Ghauspur Daho using Landsat and MODIS imagery). Band-7 of these images was density sliced and there-off vectorized to extract the stream channels. On the parallel, it was considered necessary to map the flood duration and its effect on the stream migration. Hence, for this purpose, MODIS images for the flood duration, *i.e.* July to September for the years 2003, 2005 and 2010 were downloaded.

For processing the vectorized files, a comprehensive geo-processing methodology (Fig. 3) was devised. This involved using Digital Shoreline Analysis System (DSAS) to calculate the change rates of the river bank on the west. In the first step, after the stream polygons were generated, the western shorelines were isolated for each of the Landsat images. These were then bundled together using the ArcGIS merge toolset within a geo-database. A baseline file was created as a reference point for transect generation (see Fig. 4), and transects were drawn manually to cut the entire set of river shorelines. Finally, the change rates alongside each of these transect were calculated and tabulated (Fig. 6).

On the other hand, water masks were generated for the MODIS imagery datasets. These images were then combined in GRASS-GIS and using the "r.series" module with the help of a custom Perl script, flood duration maps were generated. These maps were then cross tabulated against individual transects to calculate the relationship between observed stream migration rates and flood water residence time (Fig. 7).

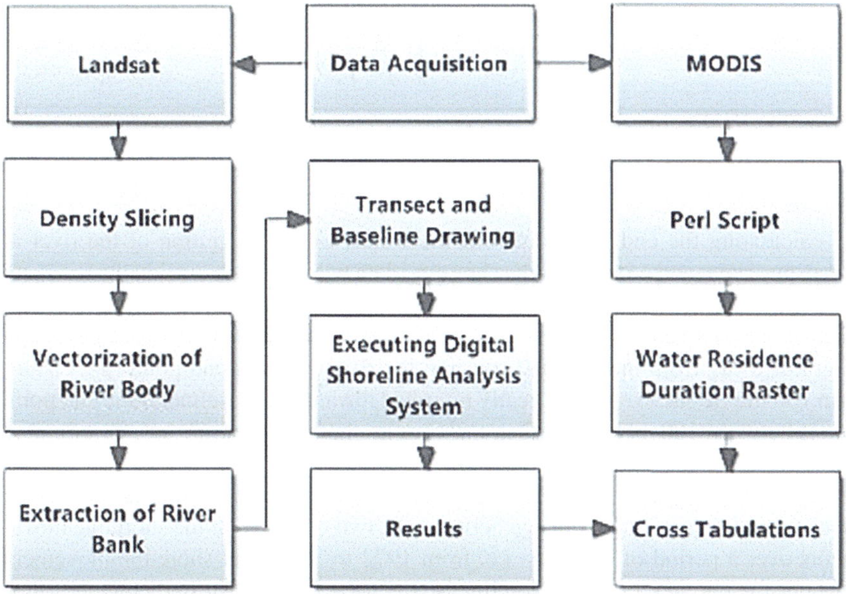

Fig. 3 Flow chart of the processing methodology used

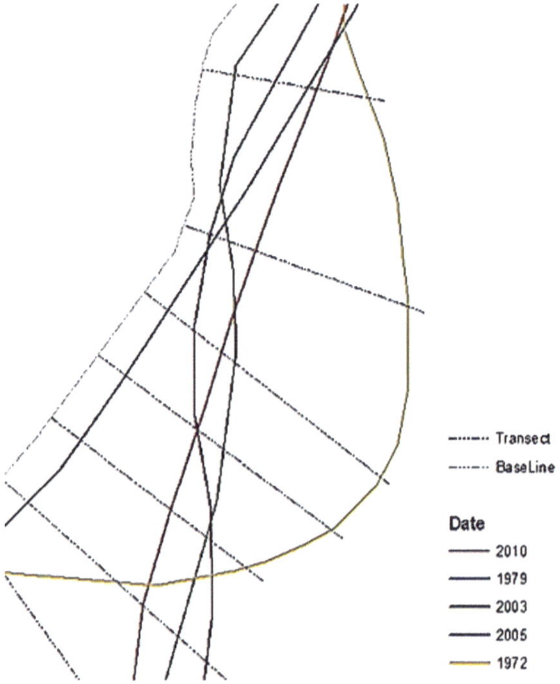

Fig. 4 An inset of the baseline and transects that were generated using the Digital Shoreline Analysis System (DSAS)

4 Results and Discussion

By calculating the end point rates for the western flank migration of the river at Ghauspur Daho and combining the shoreline data with baseline outside the river and then by using transects, the erosion rates for the western flank were visualized. If the end point rate has value less than −100, then it implies that the river has greatly migrated away from the baseline and into the river whereas end point rate greater than 100 implies that river has greatly migrated towards the baseline. If the end point rate is between 100 and −100, then this implies that the shoreline has not migrated much either away or towards the baseline (Fig. 5).

Shoreline change envelope for the western flank of the river at Ghauspur Daho gives the maximum rate of change between the two extremes of the shoreline movements over a period of 41 years, i.e. from 1972 to 2013. Net shoreline movement explains the rate of change in shoreline between the first year in study and the latest year included in the study. The end point rate is calculated by dividing the distance of shoreline movement by the time elapsed between the oldest and the most recent shoreline. The curves were plotted between transects on x-axis and rate of change on y-axis. The curves describe the erosion rates for the western flank of the river at Ghauspur Daho. The positive peaks show that at that very point the river has highest migration towards the baseline and negative peaks stand for the highest migration of western flank of river away from the baseline (Fig. 6).

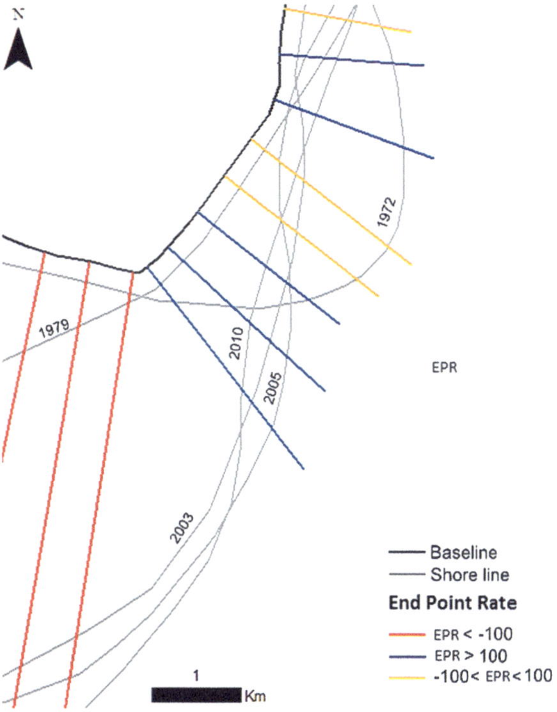

Fig. 5 An inset of baseline and transects showing different rates of erosion based on end point rates

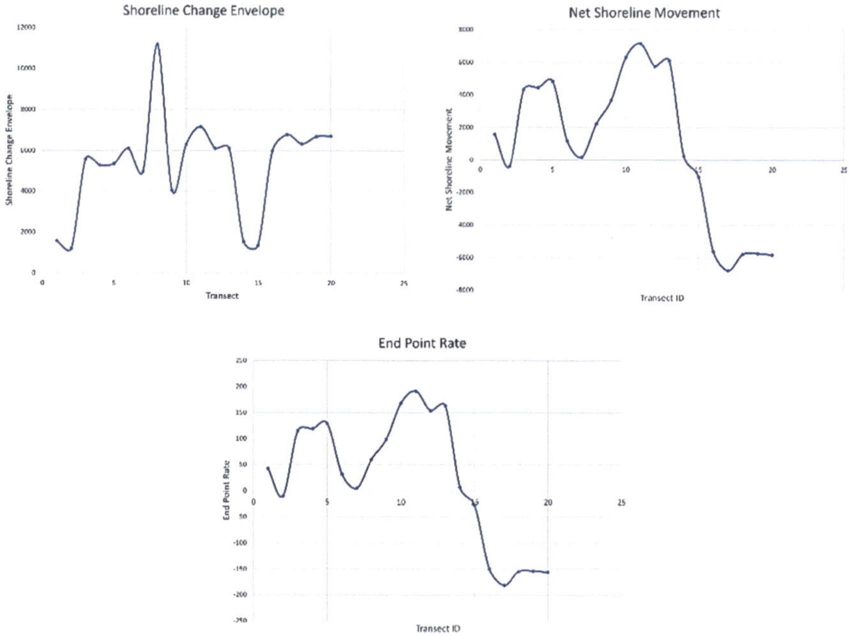

Fig. 6 Curves showing the erosion rates over the period of 41 years

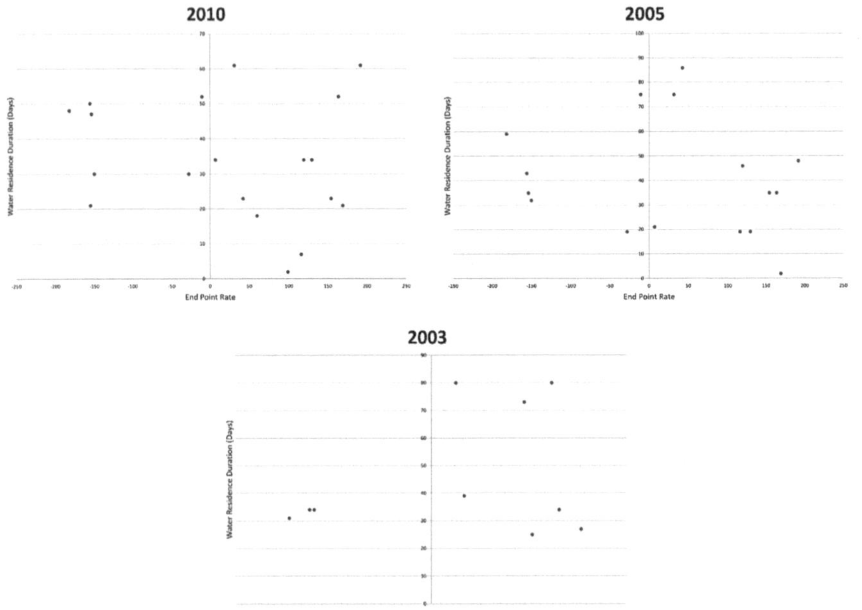

Fig. 7 Effect of flood water residence time on the shoreline erosion rates

The general relationship between the flood water residence time and the erosion rates is that if residence time of flood water increases it increases the erosion rate of the river shoreline. From the graphs plotted for flood durations in year 2003, 2005 and 2010 (Fig. 7), it is quite obvious that higher residence time of flood water increases the erosion rate and hence is the major contributor to the change in natural course of the river at Ghauspur Daho in the last decade.

5 Conclusion

The remote sensing of the shoreline migration over the past 40 years indicates that the river shows a regular avulsion in its course over a period of many years due to the deposition of sediments from the land erosion and hence causing the river bed to rise at that point. This forces the river to change its channel and find a place somewhat lower to flow through. However, this migration of the river has greatly been increased by the flooding in Indus River. Major floods such as that of 2003, 2005 and 2010 have caused much devastation. Practices for flood mitigation, e.g. storage pools (polders and levees) on the western flank of the river, have caused it to migrate towards west. Ghauspur Daho River segment is the major breach that happened on the western flank of the River Indus ~61 river km south of Guddu barrage.

The increased rate of erosion along western shoreline is most probably the outcome of flood and practices for flood mitigation. Flood water residence time affects the rate of erosion significantly. Greater is the flood water residence time, higher will be the erosion rates. The levees constructed along river western flank for disaster mitigation increased the erosion when they failed and resulted in breach formation. As during the inundation duration, the river course may be blocked by the debris from the land destruction. So due to higher flux of the river the flow of water into the western breach increased and further storing water in polders facilitated the erosion rates.

Finally, it is concluded that the practices for disaster mitigation, on the one hand, prove beneficial for the environment by protecting the areas along rivers from severe consequences of flood but, on the other hand, they are disturbing the natural evolution of the river, hence creating an imbalance in the environment. Such an imbalance is responsible for making the floodplains more vulnerable to natural hazards.

6 Recommendations

As evident from the study, disaster risk reduction measures taken in order to mitigate the negative impacts, often end in enhancing the vulnerability instead of reducing it, if not undertaken sustainably. In developing country like Pakistan, it must be ensured that disaster risk reduction measures taken are sustainable and pose no or least threat to the community so that budget is rightly managed and no sudden diversion of funds in case of disaster takes place. Also check and balance is required to maintain the activity and to ensure effective implementation of the disaster mitigation measures.

References

Chen, S.-S., Chen, L.-f, Liu, Q.-H., Li, X., & Tan, Q. (2005). Remote sensing and GIS-based integrated analysis of coastal changes and their environmental impacts in Lingding Bay, Pearl River Estuary, South China. *Ocean & Coastal Management, 48*(1), 65–83.

Dunne, T., & Leopold, L. B. (1978). *Water in environmental planning*. Macmillan.

Galloway, G. E., & Lewis, E. L. (2012). *A white paper: The need for a unified national program for floodplain management in the 21st century*. Report Prepared for the Federal Interagency Floodplain Management Task Force.

Khan, N. I., & Islam, A. (2003). Quantification of erosion patterns in the Brahmaputra-Jamuna River using geographical information system and remote sensing techniques. *Hydrological Processes, 17*(5), 959–966.

Lancaster, S. T., & Bras, R. L. (2002). A simple model of river meandering and its comparison to natural channels. *Hydrological Processes, 16*(1), 1–26.

Leopold, L. B., & Wolman, M. G. (1957). *River channel patterns: Braided, meandering, and straight*. US Government Printing Office.

Poesen, J. W. A., & Hooke, J. M. (1997). Erosion, flooding and channel management in Mediterranean environments of southern Europe. *Progress in Physical Geography, 21*(2), 157–199.

Serra-Llobet, A., Jähnig, S. C, Geist, J., Kondolf, G. M., Damm, C., Scholz, M., Lund, J., Opperman, J. J., Yarnell, S. M., Pawley, A., Shader, E., Cain, J., Zingraff-Hamed, A., Grantham, T. E., Eisenstein, W., & Schmitt, R. (2022). Restoring rivers and floodplains for habitat and flood risk reduction: Experiences in multi-benefit floodplain management from California and Germany. *Frontiers in Environmental Science, 9*, 765.

Sung, K., Jeong, H., Sangwan, N., & Yu, D. J. (2018). Effects of flood control strategies on flood resilience under sociohydrological disturbances. *Water Resources Research, 54*(4), 2661–2680.

Yang, X., Damen, M. C. J., Zuidam, V., & Robert, A. (1999). Satellite remote sensing and GIS for the analysis of channel migration changes in the active Yellow River Delta, China. *International Journal of Applied Earth Observation and Geoinformation, 1*(2), 146–157.

Assessing the Impact of Land Use/Land Cover Changes on Hydrological Components of the Jhelum River Basin Using SWAT Model

Muhammad Uzair Akhtar, Syeda Maria Zafar, Haris Lateef, Muhammad Salman Shaheer, Usama Shahid, Muhammad Sarim Azeem, and Usman Bakht

1 Introduction

Land cover and land use change (LULC) is changing water cycles and can potentially have huge impacts on Earth's water. The leading factor in causing changes to the hydrological procedures, such as runoff, ET, and water yield is land use/land cover change. The progression of the LULC essentially affects environmental components such as precipitation and temperature, which are the main drivers of the hydrological cycle. In doing so, it alters the water balance of a watershed, which includes water yield, runoff, and evaporation. It is then imperative for hydrologists, biologists and land managers to consider the practices and outcomes of changing LULC.

1.1 Techniques

Many hydrological models such as HEC-HMS, InVEST, VIC and SWAT are used to find out how land cover and land use changes affect the hydrology of the watershed. The SWAT has demonstrated its compatibility in circumstances of restricted information accessibility in hydrological research. As a result, it is an accurate indication to consider for analyzing the impact of land use shift on water resources in Pakistan's Jhelum River Basin.

M. U. Akhtar (✉) · S. M. Zafar · H. Lateef · M. S. Shaheer · U. Shahid · M. S. Azeem · U. Bakht
Institute of Geographical Information System, National University of Science and Technology NUST, Sector H-12, Islamabad, Pakistan
e-mail: makhtar.be19igis@student.nust.edu.pk

N. Shahzad (ed.), *Water and Environment for Sustainability*,
https://doi.org/10.1007/978-3-031-27280-6_5

1.2 Studies at Regional Level

Numerous pieces of research have evaluated the impacts of LULC alterations on local water supplies. For example, Li et al. (2018) observed the growth of the developed region and the decrease of the vegetation region in Jing-Jin-Ji, China which results in a rise in water yield (5%). Mango et al. (2011) organized surveys across the upper Mara Waterway in Kenya and showed that assuming forest cover was transformed to grassland, surface runoff increased by 20% but ET dropped by 2%. Also, Zhu et al. (2014) measured the effects of land cover and land use changes on the hydrometry of the Little Waterway Bowl, Tennessee. The findings highlighted a minor 3% rise in current flow, but notable spatial changes throughout the dish. Ahiablame et al. (2017) studied the effects of two prospective land uses (LULC-2055 and LULC-2090) under three reproductive situations (A1B, A2, and B1) on the stream flow of the James River watershed, USA, and also tracked that environmental land use changes would be against lead to 12–18% and 17–41% expansions in yearly current flow by the end of the twenty-first century.

1.3 Studies at the Local Level

Research on LULC alterations has been done in Kashmir (a region of the upper Jhelum River Basin) and northern Pakistan. WHO emphasized the main reasons it has supplied information on deforestation and urban expansion in the city of Islamabad from 1992 to 2012 and its ecological determinants. Likewise, Butt et al. (2015) provided detail about the simple watershed in which they highlighted a surge in colonization and a fall in vegetation. Mannan et al. (2019) have detected the growth in urban areas and agriculture as well as a decline in forest cover in the lower region of the Himalayan Mountains in Pakistan. An investigation had conducted by Iqbal and Khan (2014) in the Azad Jammu and Kashmir subdivision to observe the spatially transient land use/land cover change somewhere in the range of 1998 and 2009 and found a fall in the forest and exposed vegetation region increasing the settlement. Alam et al. (2014) investigated the LULC variations somewhere between 1992 and 2015 in the Kashmir Valley, India, found a rise in plantations, and developed regions with a simultaneous decrease in agriculture. In any case, only a single report was made over the limited area of the Indus to measure the results of LULC by employing the HEC-HMS hydrological model. Younis and Ammar (2017) established that a general variation of publication is irrelevant.

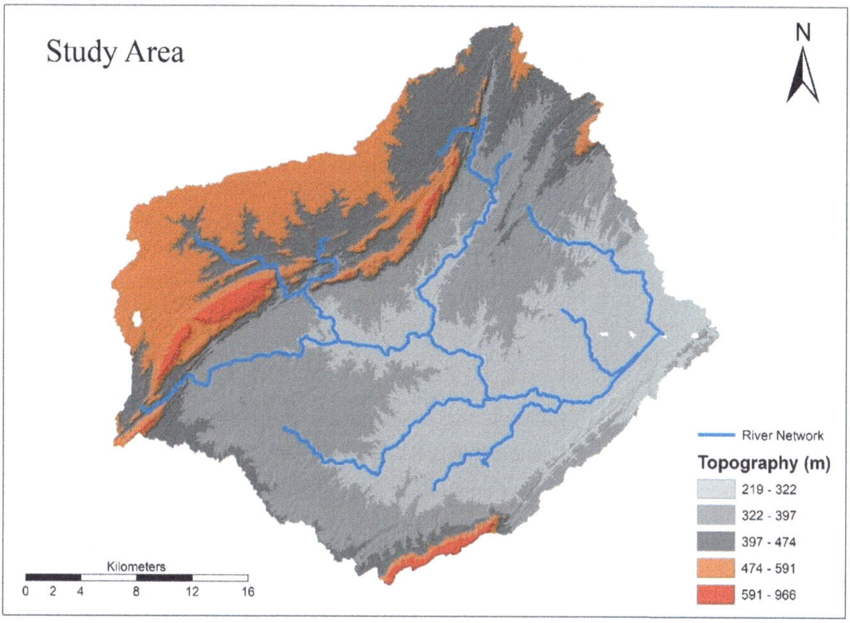

Fig. 1 Jhelum River Basin

1.4 Study Area

The Jhelum River drainage basin is situated between E 73.7259 and N 32.9409 and has a total drainage area of 982 km. Figure 1 displays the map of the concerned area. Altitudes in the basin terrain span from 219 m in the plains to 966 m in the peaks. The entire basin has an average annual rainfall of 1254.4 mm and an average annual temperature of 15.7 °C. Most of the precipitation falls from March to August. The main soil types in the basin are Lithosols, Calcaric Regosols, Eutric Cambisols and Haplic Xerosols. The basin area is enclosed by varied land covers, such as Barren Land, Natural Vegetation, Farmland, and Urban and Built-Up areas.

2 Study Design

2.1 Data Description

The day-to-day perceived rainfall, temperature, and relative humidity statistics of the Jhelum weather station for the years 2001, 2010 and 2020 were collected from NASA POWER (https://power.larc.nasa.gov/data-access-viewer/). Landsat Imagery for the years 2001, 2010 and 2020 was collected from USGS (EarthExplorer). Table 1

Table 1 Properties of Landsat images

Year	Satellite	Path/row	Resolution	Acquisition (dd/mm/yyyy)	CloudCover
2001	Landsat8	148/37, 149/36-37, 150/36–37	30 m	20/1/2001	<10%
2010	Landsat8	148/37, 149/36-37, 150/36–37	30 m	14/5/2010	<10%
2020	Landsat8	148/37, 149/36-37, 150/36–37	30 m	09/3/2020	<10%

Table 2 Soil class representing the basin

Soil classes	Area (%)
Lithosols	88.62
Calcaric Regosols	2.82
Eutric Cambisols	2.80
Haplic Xerosols	5.76

describes the properties of Landsat images obtained. The soil data were collected from the website of FAO. Table 2 gives details on the various soil classes and the area they took. The importance of soil data was to help in HRU (Hydrological Response Unit) analysis and definition. It was also used to create the soil table in the SWAT model.

Digital Elevation Model (DEM) downloaded from DIVA-GIS. Resampling was applied to give a 30 m resolution to the DEM. The DEM was employed for watershed delineation, along with the determination of streams, outflows and inlets, as well as the computation of the characteristics for sub-basins. The DEM is also used to determine the slope of the study area, hence, determining the seasonal offsets of the area concerning water recharge and the nature of land.

2.2 *LULC Classification*

Figure 2 explains the steps we took to perform LULC classification and change. The first step was to remove distortions from the satellite images and perform pre-processing. ERDAS software was used for atmospheric, geometric and radiometric corrections for images. Image mosaicking, subsetting and supervised classification were performed in ArcGIS. LULC was classified into 4 classes namely Barren Land, Natural Vegetation, Croplands and Urban and Built-Up. Table 3 lays out a detailed illustration of distinct LULC classes. The accuracy assessment was performed for the Jhelum River Basin in 2001, 2010 and 2020. User accuracy was calculated based on classification for 4 different classes. For producer accuracy reference image

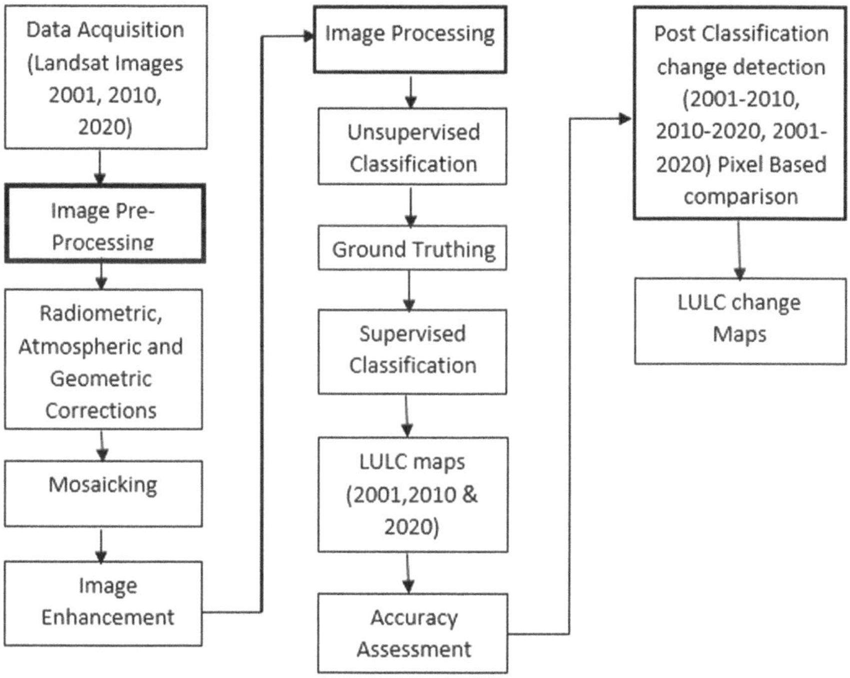

Fig. 2 Flowchart of methodology for LULC

Table 3 LULC classes

Class code	LULC class	Description
1	Barren land	This class includes mountainous rangelands, barren land, dry stream channels and sparsely vegetated area
2	Natural vegetation	Evergreen, grasslands and deciduous forests, etc.
3	Waters	Rivers, lakes and ponds
4	Croplands	Wheat, rice, fodder crops and vegetables
5	Urban and built-up	Residential areas, commercial areas, roads and industrial zones

from Google Earth Pro was used. The overall accuracy and Kappa Coefficient were calculated by generating a 4 by 4 pair-wise comparison matrix.

$$\text{User Accuracy} = \frac{\text{Number of correctly classified pixels in each category}}{\text{Total number of classified pixels in that category (the row total)}} \times 100 \quad (1)$$

$$\text{Producer Accuracy} = \frac{\text{Number of correctly classified pixels in each category}}{\text{Total number of reference pixels in that category (the column total)}} \times 100 \quad (2)$$

$$\text{Overall Accuracy} = \frac{\text{Total number of correctly classified pixels (diagonal)}}{\text{Total number of reference pixels}} \times 100 \quad (3)$$

$$\text{Kappa Coefficent} = \frac{(\text{TS} \times \text{TS}) - \sum (\text{column total} \times \text{row total})}{\text{TS}^2 - \sum (\text{column total} - \text{row total})} \times 100 \quad (4)$$

2.3 SWAT Hydrological Model

Figure 3 is a summary on the concept behind the SWAT, the various stages and input data needed in running a successful model. The stages involve watershed delineation which make use of the DEM for the generation of streams, their inlets and outlets, flow direction, accumulation and creation of basins and sub-basins. The HRUs make use of LULC, soil raster data, and slope as its input data. It then generates sub-units for the watershed based on their similarities with respect to the soil, LULC and slope. The next stage after the HRUs definitions is the write-input tables. This section of the model makes use of climate data and the creation of tables needed for your results from the watershed delineation all through to the HRUs and the climate data as a whole. Then the final stage is the simulation stage, which involves the setting of the date range for simulating and results in parameters of interest being modeled. Water balance is calculated with the help of the equation:

$$\text{SW}_t = \text{SW}_0 + \sum_{i=1}^{t} \left(R_{\text{day}} - Q_{\text{surf}} - E_{\text{a}} - W_{\text{seep}} - Q_{\text{gw}} \right), \quad (5)$$

where the final soil water contents (mm) is denoted by SW_t and initial soil water content (mm) is denoted by SW_0; t is the time in days; R_{day} is the amount of precipitation on day i (mm); Q_{surf} is the count of surface runoff on day i (mm); E_{a} is the amount of ET (mm); W_{seep} is the quantity of water that enters the vadose zone from soil profile (mm); Q_{gw} is the amount of base flow on day i (mm). The water yield was calculated with the help of the equation:

$$Q_{\text{yld}} = Q_{\text{srf}} + Q_{\text{lat}} + Q_{\text{gw}} - T_{\text{loss}}, \quad (6)$$

where the quantity of water yield (mm) is denoted by Q_{yld}; the surface runoff (mm) is denoted by Q_{srf}; the quantity of water provided by lateral current (mm) is denoted by

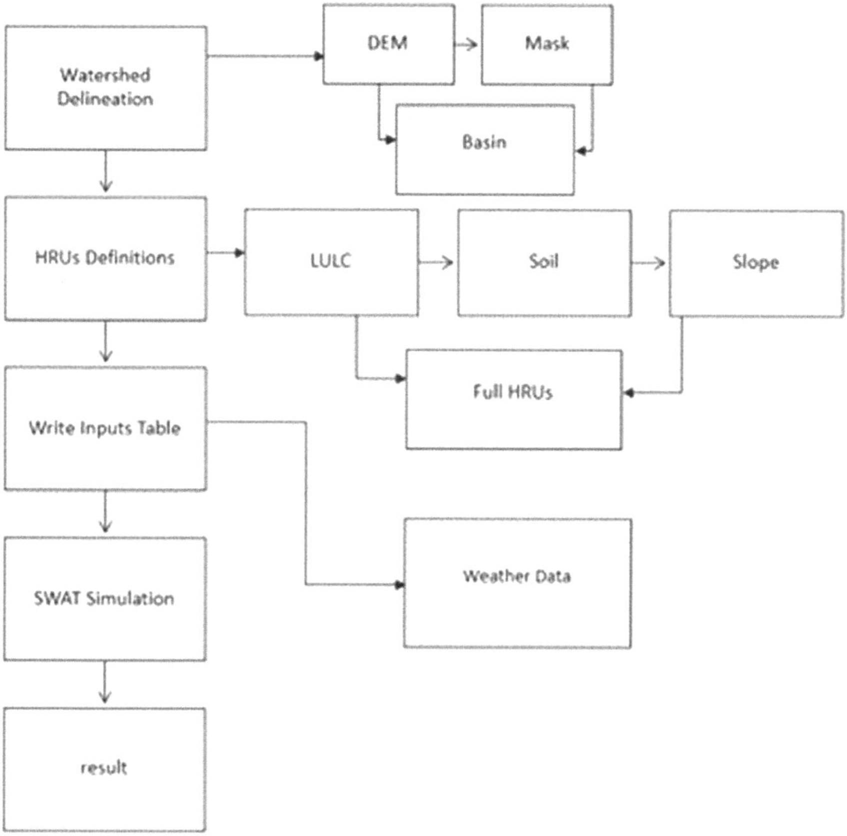

Fig. 3 Conceptual framework of the SWAT model setup

Q_{lat}; the contribution of stream flow (mm) is denoted by Q_{gw} whereas T_{loss} represents the loss of water through transmission process (mm). For soil and slope, a threshold of 5% has been used for this study and a threshold of 10% was used for land use and the Jhelum River Basin was divided into 23 sub-basins and 67 HRUs. To calculate evapotranspiration, the Hargreaves method has been used since the data on air velocity and solar energy were not accessible for the simulation duration. The SWAT model was running separately for 2001, 2010 and 2020 data. Figure 4 is the map of the Jhelum River Basin alongside its sub-basins and flow gauges.

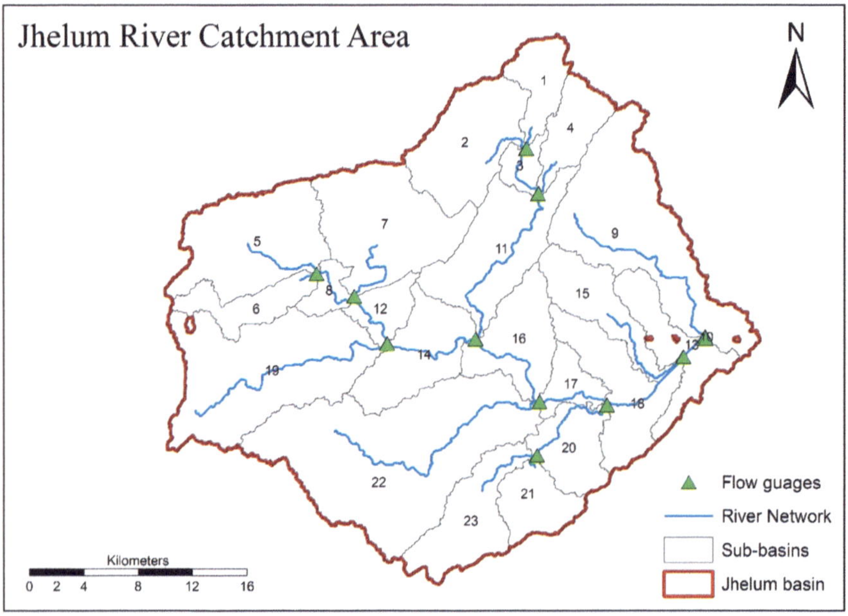

Fig. 4 River basin in the concerned area and sub-basins which have been generated during delineation

3 Results and Discussion

3.1 Land Cover and Land Use Classification

Table 4 depicts the results of the accuracy assessment of 4 classes in terms of user's accuracy (UA), producer's accuracy (PA), overall accuracy (OA) and Kappa coefficient (KC) for 2001, 2010 and 2020. Figure 5 depicts the geographical pattern of LULC for the years 2001, 2010 and 2020.

Table 5 displays the LULC matrix for the period from 2001 to 2020, and Table 6 shows the quantitative alteration that has been taking place in LULC during the past 19 years. Significant variations can be noticed in the natural vegetation (decrement) and barren land (increment) classes. This is the result of deforestation in the basin. Figure 6 illustrates the changes that have been taking place within each class during distinct intervals (2001–2010, 2010–2020, 2001–2020). It can be clearly observed that major changes occurred between 2001 and 2020.

Table 4 LULC maps accuracies

LULC class	2001				2010				2020			
	UA (%)	PA (%)	OA (%)	Kc (%)	UA (%)	PA (%)	OA (%)	Kc (%)	UA (%)	PA (%)	OA (%)	Kc (%)
Barren land	100	28.6	73	61.3	60	42.9	76.7	65.1	100	81.8	86.7	81.8
Natural vegetation	61.5	80			80	85.7			77.8	87.5		
Croplands	76.9	100			75	100			75	100		
Urban and built-up	100	66.7			100	66.7			100	80		

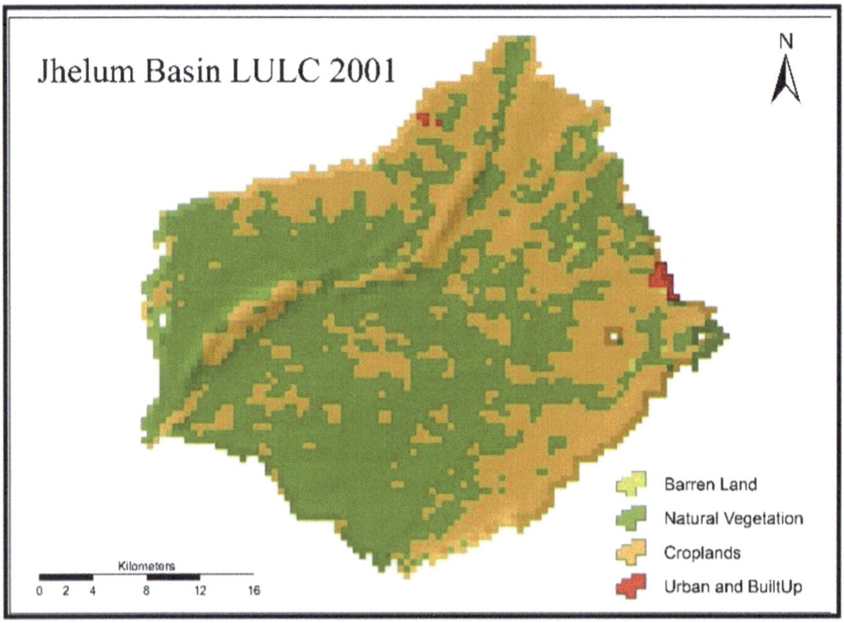

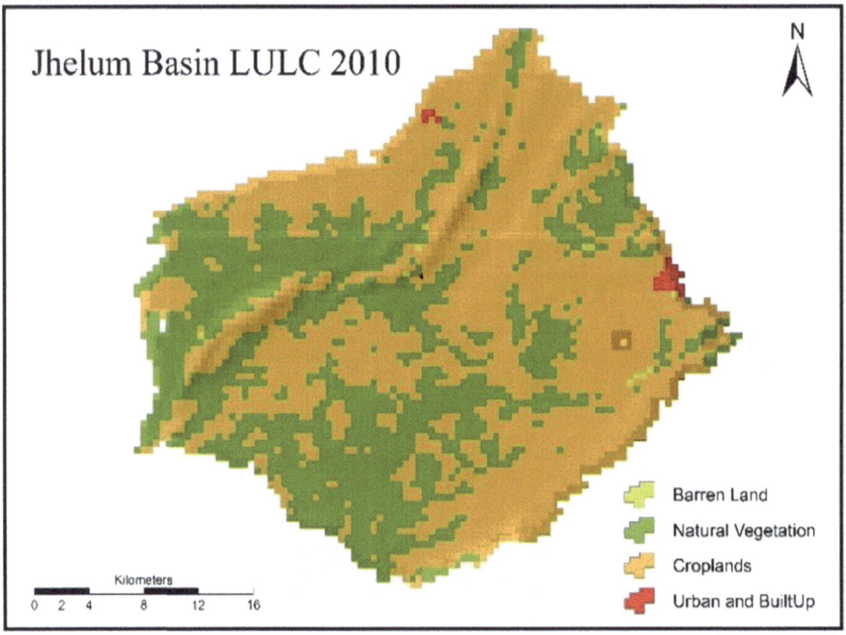

Fig. 5 Maps for the years, 2001, 2010, 2020, for LULC of Jhelum River Basin

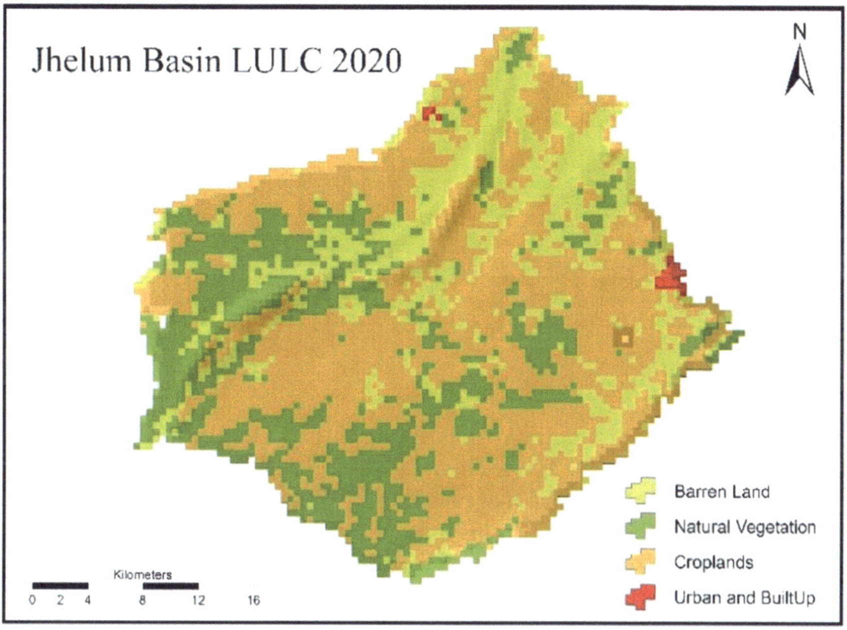

Fig. 5 (continued)

Table 5 LULC (Km^2) matrix for the period of 2001–2020

		LULC 2020				
		Barren land	Natural vegetation	Croplands	Urban and built-up	Total
LULC 2001	Barren land	0.66	0.22	0.33	0.11	1.32
	Natural vegetation	119.33	221.89	232.86	0.10	574.18
	Croplands	81.48	18.52	302.22	0.73	402.95
	Urban and built-up	0.27	0.09	0.12	3.07	3.55
	Total	201.74	240.72	535.53	4.01	982.0

3.2 *SWAT Calibration and Validation*

3.2.1 Land Use Change Effects

To research the effects of LULC modifications on the hydrodynamic elements in the Jhelum River Basin, in the swat model land use from three different time intervals has been used, i.e., (2001, 2010 and 2020) while all other inputs except climate data, such as, e.g., B. Precipitation, temperature and relative humidity were held identical.

Table 6 Area statistics and changes in LULC for the period of 2001–2020

LULC Class	2001	2010		2020		Change 2001–2020	
	Area (Km^2)	Area (%)	Area (Km^2)	Area (%)	Area (Km^2)	Area (%)	Area (Km^2)	Area (%)
Barren land	1.32	0.15	2.23	0.26	201.74	20.82	+200.42	+20.76
Natural vegetation	574.18	57.96	387.32	39.21	240.72	24.74	−333.46	−33.22
Croplands	402.95	41.51	599.35	60.12	535.53	54.03	+132.58	+12.52
Urban and built-up	3.55	0.39	3.63	0.41	4.01	0.41	+0.46	+0.02

The evaluation included runoff, WY and ET under LULC outline of different time intervals (2001/2010/2020). The results of surface runoff, water yield and ET are given in Table 7. It can be seen that runoff and WY are decreasing over 19 years. On the other hand, ET increases. This increase is partly due to the decrease in the region of natural vegetation and the increase in arable land and wasteland. Due to the rise in the arable area and the decrease in natural vegetation, the rate of transpiration and infiltration has increased drastically which causes a major rise in ET. The net annual water yield is the difference between precipitation that fell in the water shell and total evapotranspiration, the decrease was observed due to the fact that total evapotranspiration was greater than the precipitation recorded over the years 2001, 2010 and 2020. The decrease in surface runoff was observed due to a decrease in natural vegetation causing a temperature increase and a decline in precipitation; moreover, the rise in cropland areas also promoted the decrease.

3.2.2 Hydro-Climatic Changes

i. Rainfall
   To realize the variability of rainfall to detect the possible climate change in the basin, we collected monthly rainfall data and plotted a graph for it in Excel to show the trend. Figure 7 shows the rainfall trend between the periods of 2001, 2010 and 2020.

ii. LULC effects on inflow parameters
   The inflow parameter considered here is the groundwater recharge value. This experienced changes as a result of changes in the LULCs. The change was determined by running the SWAT model twice in 2010 and 2020. LULC maps on separate accounts with the soil data being kept as constant variable and weather data also being added which we collected. The recharge to deep aquifers increases from 2.1 mm in 2010 to 10.02 mm in 2020. This is due to the fact that agricultural activities have increased thus the construction of more wells, infiltration basins

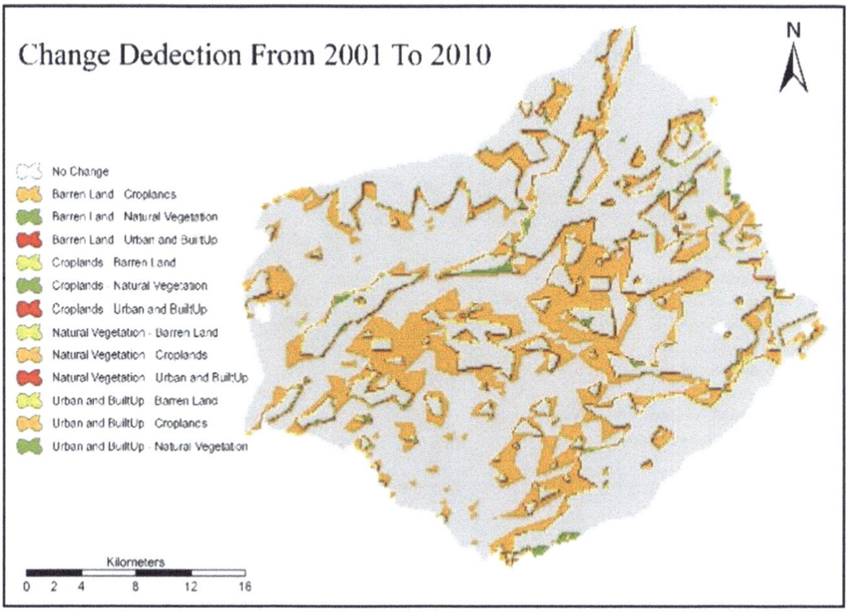

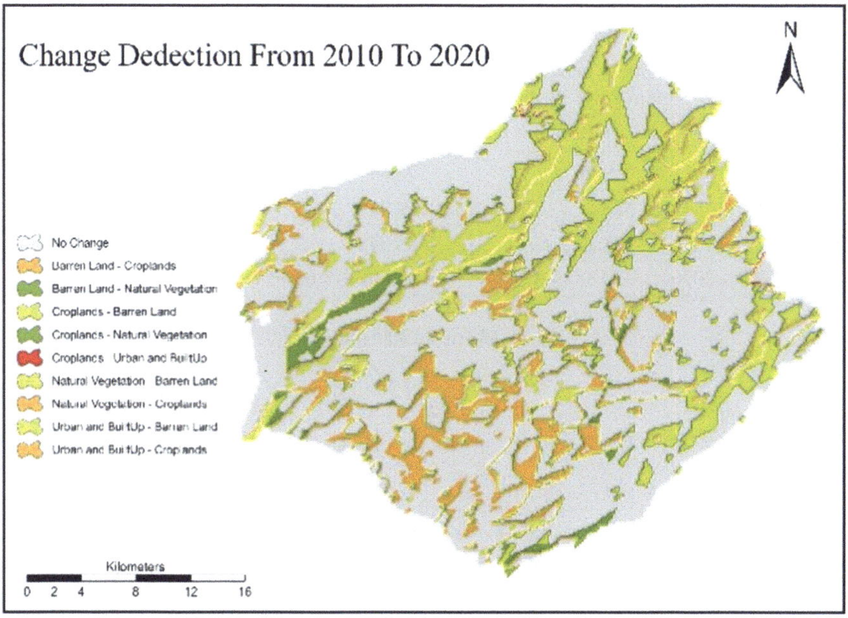

Fig. 6 Change detection of land use/land cover of the Basin of Jhelum River during the intervals of 2001–2010, 2010–2020 and 2001–2020

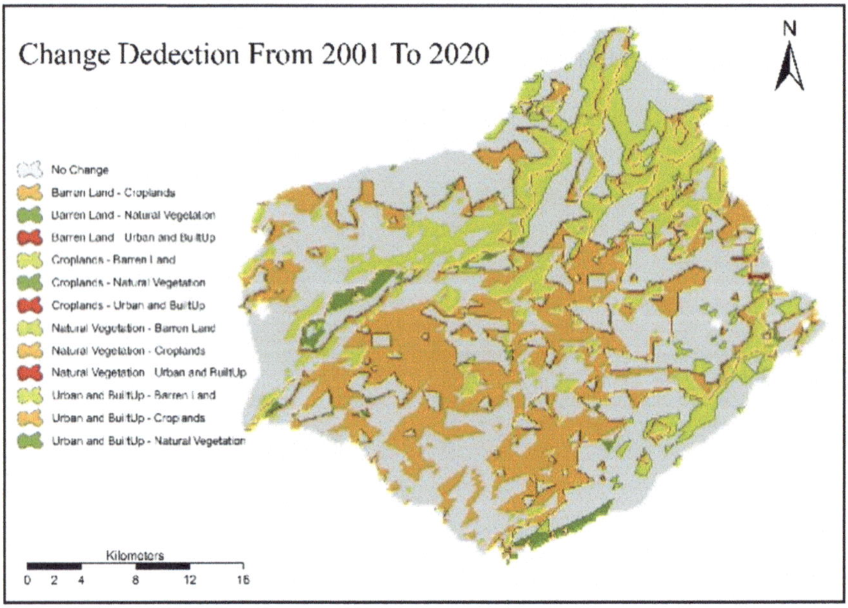

Fig. 6 (continued)

Table 7 Correlation of mean annual hydrological components using the three historical LULC in SWAT model over Jhelum River Basin

LULC	Surface runoff (mm)	Evapotranspiration (mm)	Water yield (mm)
LULC 2001	539.03	593.84	927.67
LULC 2010	527.21	649.91	920.58
LULC 2020	508.15	686.64	910.87
Changes 2001–2020	−30.88 (−5.72%)	+92.80 (+15.62%)	−16.80 (−1.81%)

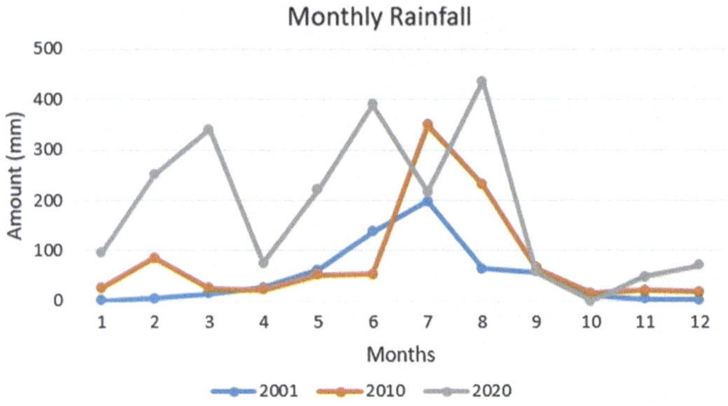

Fig. 7 Graph showing rainfall trend

and ponds helps redirect the water across the land surface. Figure 8 shows the hydrology parameters for 2010 and 2020.

3.2.3 Effects of Landscape

i. Sediment
Sediment loss from the landscape is dependent upon many factors. Sediment overestimation in SWAT is most commonly due to inadequate biomass production. This often occurs on specific land uses. SWAT also modifies sediments to account for in-stream deposition and erosion of stream banks and channels. Often there is little or no measured data to differentiate between upland sediment and in-stream sediment changes. Streams may be either a net source of sediment or a sink. In-stream sediment modification is impacted by physical channel characteristics (slope, width, depth, channel cover, and substrate characteristics) and the quantity of sediment and ow from upstream. Figure 9 shows a difference in the average upland sediment yield and stream sediment change.
ii. Plant Growth
Proper plant growth is key to accurate runoff and sediment predictions. Problems in plant growth are often related to excessive stress due to temperature or the lack of water/nutrients. Phosphorous (P) and Nitrogen (N) are vital for plant growth. Figure 10 shows how the plant growth varied among the three different periods of 2001, 2010 and 2020. Best plant growth was found in 2010 due to excess intake of P and N. This can be due to the fact that the vegetation area is greater in 2010. Table 8 shows the summary of the mean Annual Hydro-Climatic Change Analysis.

4 Conclusion

This study examines how human activity and global warming have affected the hydrodynamic elements of the river Jhelum basin. The results indicated that the Basin of river Jhelum had gone through notable changes in LULC over 19 years. Four crucial LULC classes have been recognized, including barren land, arable land, natural vegetation, and urban and built-up. Among these land uses, the barren land and arable land classes showed a noteworthy rise in the change period of 2001–2020 but all of this change has come at the cost of natural vegetation. Significant gains and losses were observed in all classes except Urban and Built-Up. The main root of the rise in fallow land was deforestation in the Jhelum River Basin to make more room for agricultural and urban activities.

The SWAT model was then run separately for three different years and it confirmed that changes in climate and LULC affects the health of a watershed. In the case of this study, the hydrological components included groundwater recharge, surface runoff,

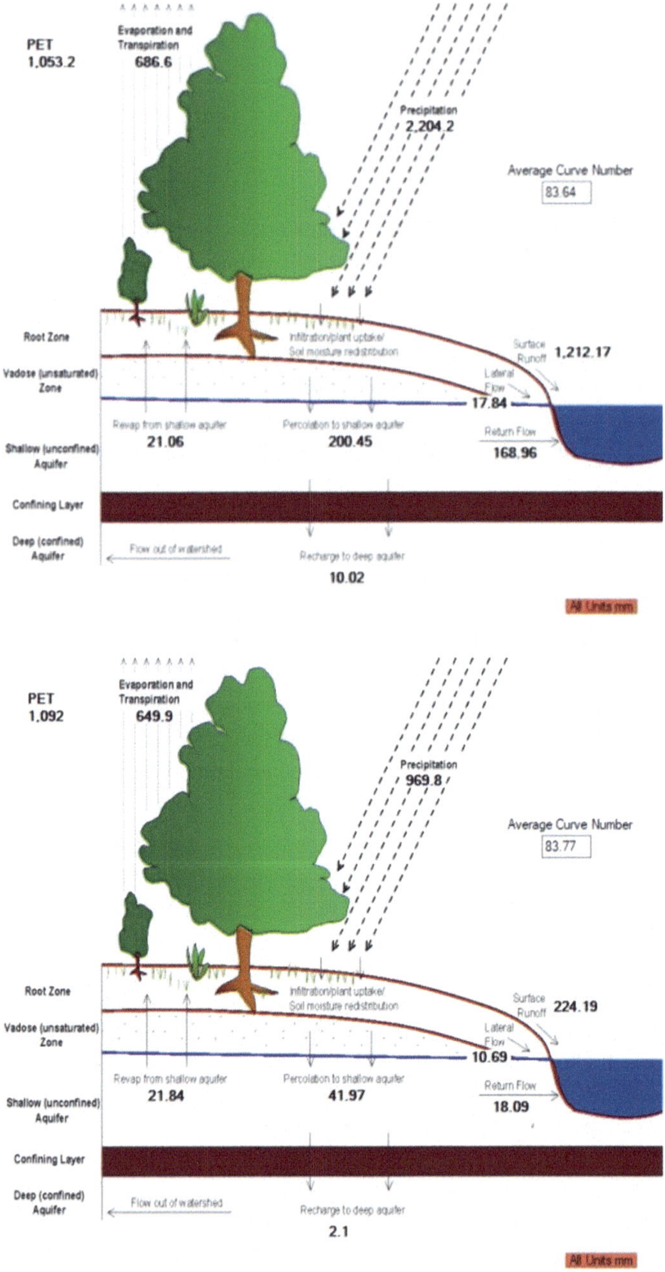

Fig. 8 Changes in hydrological components (2020 and 2010)

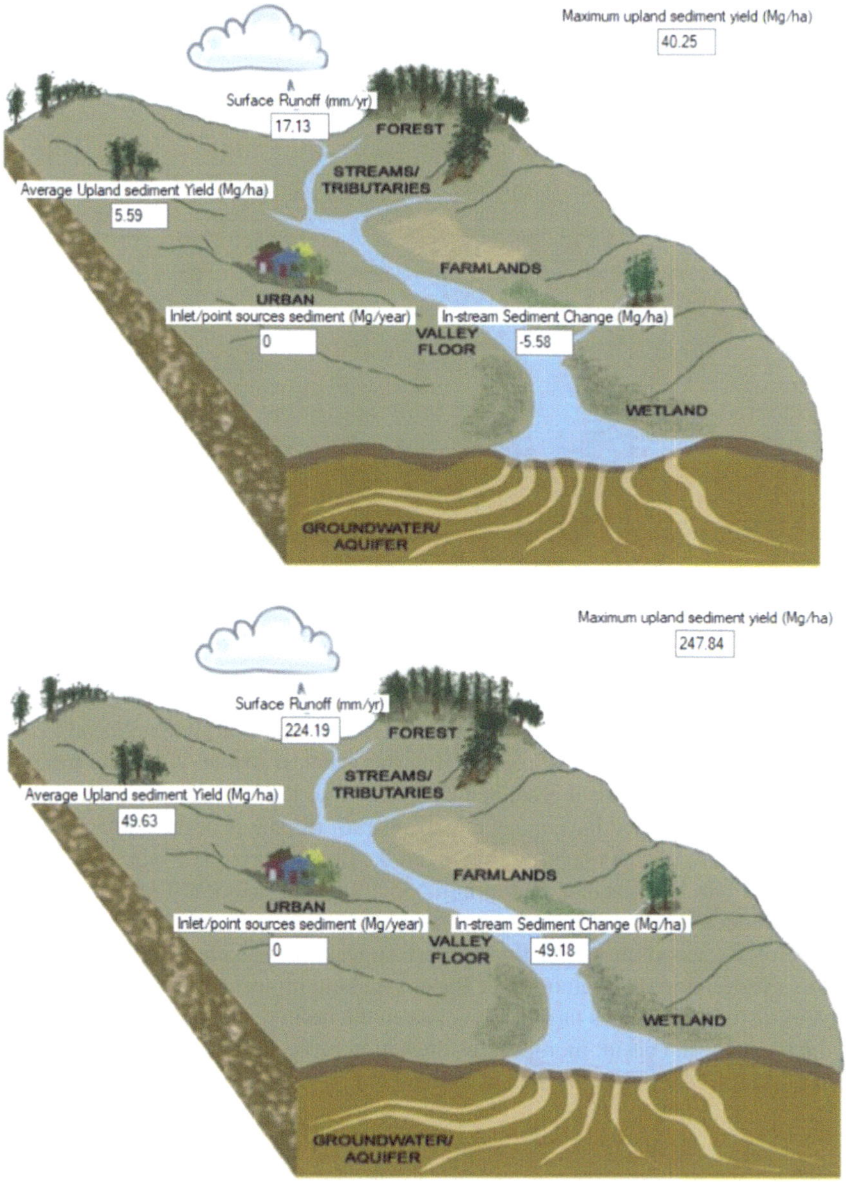

Fig. 9 Difference in the average upland sediment yield and stream sediment change

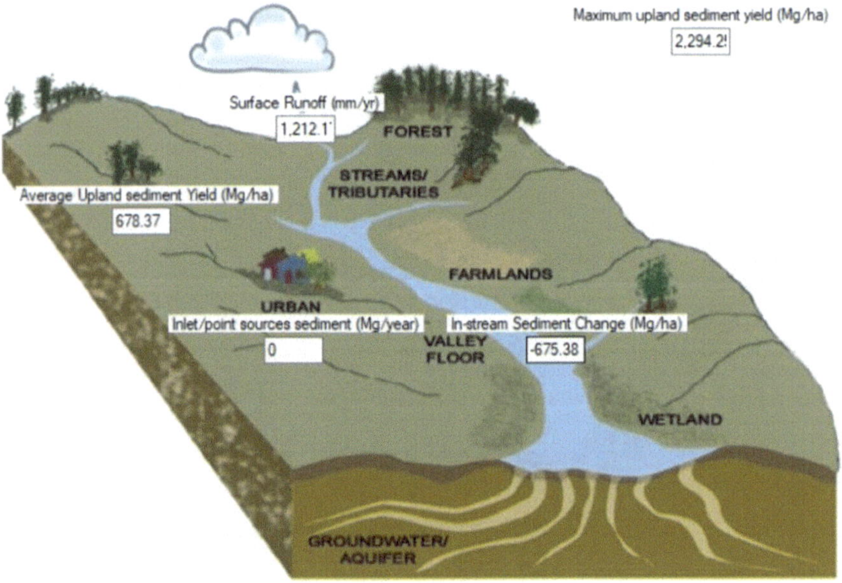

Fig. 9 (continued)

water yield and evapotranspiration. Our analysis indicates that evapotranspiration increased extremely in the basin area. Since high evapotranspiration will lead to low drainage density and effective reduction in the water available within the basin in the near future, as has been predicted by Bates et al. (2008) that, these changes in land cover/land use will have many effects on various phases of the human society, from agriculture to water supply. Due to a reduction in forest and vegetation areas, an expansion in bare ground, and farming practices, surface runoff and water yield are also reduced.

Therefore, according to our findings, we suggest the stakeholders and administrators overlooking the Jhelum River Basin area help protect the basin from anthropogenic activities, such as farming too close to the basin to balance the hydrological components of the basin; moreover, afforestation projects should be encouraged to start to prevent the rise in the area of barren land.

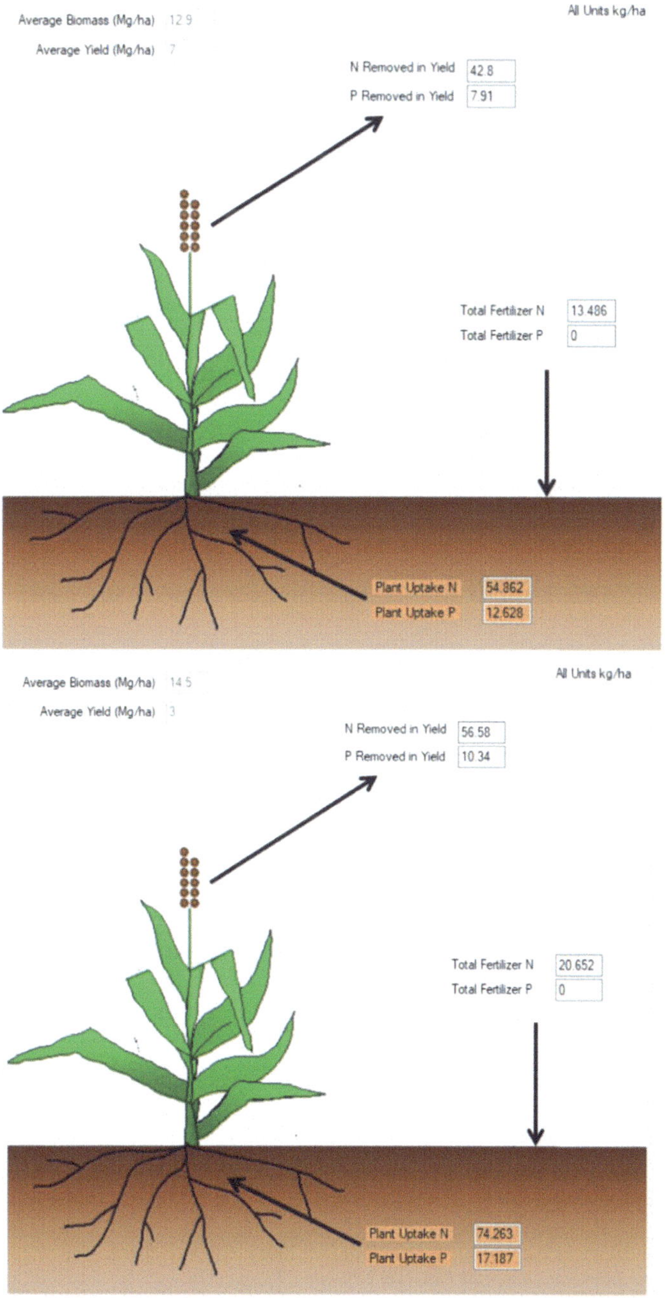

Fig. 10 Variation of plant growth in the three different periods of 2001, 2010 and 2020

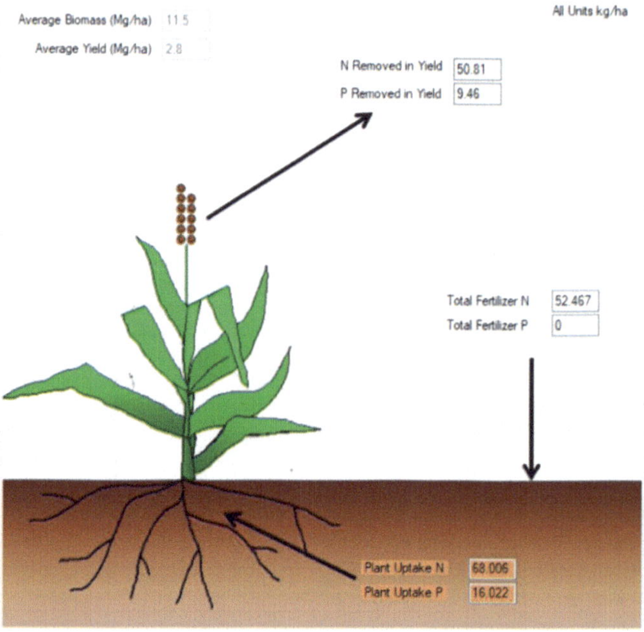

Fig. 10 (continued)

Table 8 Summary of mean annual hydro-climatic change analysis

Year	Mean annual temperature (C)	Mean annual relative humidity (mm)	Sediment (Mg/ha)	Plant uptake N (Kg/ha)	Plant uptake P (Kg/ha)
2001	22.18	34.82	40.25	54.86	12.63
2010	24.77	45.14	247.84	74.26	17.19
2020	25.23	62.83	294.25	68.01	16.02
Changes 2001–2020	3.05	28.01	254	13.15	3.39

References

Ahiablame, L., Sinha, T., Paul, M., Ji, J., & Rajib, A. (2017). Stream flow response to potential land use and climate changes in the James River watershed, Upper Midwest United States. *Journal of Hydrology: Regional Studies, 14*, 150–166.

Bates, B., Kundzewicz, Z., Wu, S., & Palutikof, J. (2008). Climate change and water. *Technical Paper of the Intergovernmental Panel on Climate Change, 210.*

Butt, A., Shabbir, R., Aziz, N., & Ahmad, S. S. (2015). Land cover classification and change detection analysis of Rawal watershed using remote sensing data. *Journal of Biodiversity and Environmental Sciences*, *6*(1), 2222–3045.

Iqbal, M. F., & Khan, I. A. (2014). Spatiotemporal land use land cover change analysis and erosion risk mapping of Azad Jammu and Kashmir, Pakistan. *The Egyptian Journal of Remote Sensing and Space Science, 17*(2), 209–229.

Li, S., Hong Yang, H., Lacay, M., Liu, J., & Lei, G. (2018). Impacts of land-use and land-cover changes on water yield: A case study in Jing-Jin-Ji, China. *Sustainability, 10*(4), 960.

Mango, L. M., Melesse, A. M., McClain, M. E., Gann, D., & Setegn, S. (2011). Land use and climate change impacts on the hydrology of the Upper Mara River Basin, Kenya: Results of a modeling study to support better resource management. *Hydrology and Earth System Sciences, 15*(7), 2245.

Mannan, A., Liu, J., Zhongke, F., & Khan, T. U. (2019). Application of land-use/land cover changes in monitoring and projecting forest biomass carbon loss in Pakistan. *Global Ecology and Conservation, 17*, e00535.

Younis, S. M. Z., & Ammar, A. (2017). Quantification of impact of changes in land use-land cover on hydrology in the upper Indus Basin, Pakistan. *The Egyptian Journal of Remote Sensing and Space Science, 21*(3), 255–263.

Zhu, Q., Nie, X., Zhou, X., Liao, K., & Li, H. (2014). Soil moisture response to rainfall at different topographic positions along a mixed land-use hillslope. *Catena, 119*, 61–70.

The Climatic Benefit of Perennial Rice Cropping System: A Case Study in West Java, Central Java, and East Java

Muhamad Khairulbahri and Adi Rivaldo

1 Introduction

There is an increasing awareness of the impacts of climate change on rice production, especially, in Asian countries. There are two main reasons for this trend. The first reason is the important contribution of Asian rice farming to employment and economic growth in Asia (Bandumula, 2018; Chauhan et al., 2017). The second one is Asia the world largest rice producer and rice consumer (Chauhan et al., 2017). In turn, possible threats to Asian rice production can affect a large fraction of the world population in terms of food security, livelihoods, and economic growth (Chauhan et al., 2017).

Among Asian countries, Indonesia is the third-largest rice producer and the third-largest rice consumer in the world (Setyanto et al., 2018). As such, Indonesia is an interesting place to understand the impacts of climate factors on Asian rice production. It is also acknowledged that Indonesian temperature, as well as Asian temperature, is relatively close to rice critical temperature, leaving Indonesia's rice production in danger under changing climate (Chauhan et al., 2017; Lobell & Gourdji, 2012; Lobell et al., 2008, 2011; Serraj & Pingali, 2018). Owing to this, some scholars have investigated several issues of Indonesian rice production. For instance, an existing study (Naylor et al., 2007) discussed the possible impacts of climate change on rice production in West Java and Bali. Other existing studies (Falcon et al., 2004; Naylor & Mastrandrea, 2009) estimated the impacts of El-Niño Southern Oscillation (ENSO) on rice production.

M. Khairulbahri (✉)
Bandung Institute of Technology, Jl Ganesha No. 10, Bandung, West Java 40132, Indonesia
e-mail: databahri@yahoo.com

A. Rivaldo
The Center for Climate Change Information, Meteorological, Climatological, and Geophysical Agency (BMKG), Jl. Angkasa I No. 2 Kemayoran, Jakarta Pusat 10610, Indonesia

N. Shahzad (ed.), *Water and Environment for Sustainability*,
https://doi.org/10.1007/978-3-031-27280-6_6

Other studies also discussed significant issues such as the impacts of climate on Indonesian rice yield (Amalo et al., 2017; Kinose & Masutomi, 2019; Kuswanto et al., 2018; Yuliawan & Handoko, 2016), modeling the rice production (Karim et al., 2019; Utami et al., 2019), the impacts of land-use change on rice production (Makbul et al., 2019; WorldBank, 2008), and adaptation of the rice farming to climate or seawater intrusion (Bahri, 2017; Rondhi et al., 2019; Sembiring et al., 2020; Sutrisno & Setyono, 2017; Utami et al., 2018).

Existing studies arguably provide useful information to investigate possible threats to Indonesian rice production. However, existing studies that analyzed the impacts of climate on Indonesian rice yield are very limited and none of the existing studies have analyzed the impacts of climate factors on three main rice provinces including West Java, Central Java, and East Java. Since the three Java provinces are responsible for half of the Indonesian rice production (BPS, 2020), this study specifically aims to investigate the impacts of climate factors on rice yield in the three main rice provinces.

Besides the aforementioned issues, Indonesian rice studies also discussed the perennial rice cropping system, so-called the salibu. This is important as most rice farmers have suffered from high-production costs such as labor and pesticide expenses (Budianti, 2021; Fitri et al., 2019; Marpaung et al., 2022; Yamaoka et al., 2017). Those studies explained that the perennial rice cropping system (the salibu) can increase rice yield, decrease short-duration growth, and decrease production cost (Budianti, 2021; Fitri et al., 2019; Marpaung et al., 2022; Yamaoka et al., 2017).

While several studies (Bahri, 2017; Khairulbahri, 2021; Welch et al., 2010) claimed that high temperatures associated with climate change may decrease Indonesian rice yield, the salibu method leads to higher rice yield recently (Budianti, 2021; Marpaung et al., 2022; Paiman, 2021). So, this study also aims to assess whether the salibu method enables rice farmers to escape from the negative impacts of high temperatures. In doing so, after the impacts of climate factors in West Java, Central Java, and East Java are assessed, the climatic benefit of the salibu is examined. In this study, the climatic benefit occurs once the salibu method can negate the negative impacts of climate factors on rice yield.

In fulfilling the research aim, this study first introduces relevant studies and the research aim. Next, the types of collected data and used methods in this study are described. Following this, this study discusses the results and explains important findings. In the end, some concluding remarks are described accordingly.

2 Materials and Methods

2.1 Three Indonesian Provinces: West Java, Central Java, and East Java

These three provinces are situated in Java, the most populated island in Indonesia. Together, these provinces are the main buffer rice stock in Indonesia (BPS Jawa Tengah, 2021; BPS Jawa Timur, 2021). The main buffer stock means these provinces provide rice not only for their local population but also to export their rice to their neighboring provinces.

West Java lies between 5°50′–7°50′ South Latitude and 104°48′–108°48′ East longitude (BPS Jawa Barat, 2021) and its capital is Bandung, so-called Paris van Java. West Java consists of 18 regencies and 9 cities with a total area and population density of about 35,377 km^2 and 1,365 persons/km^2, respectively (BPS Jawa Barat, 2021).

West Java has average annual rainfall between 2,000–4,000 mm/year, making this province the highest annual rainfall compared to the other Java provinces. Agriculture contributes to about 20% and 9% of the total employment and the economic output, respectively (BPS Jawa Barat, 2021) (Fig. 1).

Central Java geographically stretches along the equator between 5°40' to 8°30' South Latitude and 108°30′ to 111°30′ East Longitude (BPS Jawa Tengah, 2021). Its capital is Semarang with a total area and population density of about 32,801 km^{2}and 1,113 person/km^2, respectively (BPS Jawa Tengah, 2021). Central Java has an annual rainfall of about 2,500 mm/year, leading to, at least, three rice growing seasons throughout the year. Agriculture contributes to 12.5 and 29% of the total economic output and the total employment, respectively (BPS Jawa Tengah, 2021).

Fig. 1 Indonesia and three main Java provinces (West Java, Central Java, and East Java). *Source* Google Maps (n.d.)

East Java is the easternmost Java province and lies between 7.12 'South Latitude—8.48' South Latitude and between 111.0 'East Longitude—114.4' East Longitude (BPS Jawa Timur, 2021). Its capital is Surabaya, the Indonesian second largest city after Jakarta. East Java has a total area and population density of about 47,800 km^2 and 851 persons/km^2, respectively (BPS Jawa Timur, 2021). The annual rainfall is about 2,800 mm/year, and the average temperature in 2020 was 27.3 °C. Moreover, agriculture contributes to 35% and 20% of the total employment and the total economic output, respectively (BPS Jawa Timur, 2021).

Based on previous paragraphs, it can be concluded that agriculture has an important contribution to employment and economic growth in given Java provinces. So, this study can give important insights for policymakers in supporting important agricultural contributions to employment and economic output.

2.2 The Salibu and Ratoon Methods

As seen in Table 1, rice growth can be categorized into three important stages including the vegetative, the reproductive, and the ripening stages (Yoshida, 1981).The length of different stages of rice growth is varied between long- and short-duration growth varieties. But the main difference is in the length of the vegetative stage (Moldenhauer & Slaton, 2001). While the length of the vegetative stage is about 30 days for short-duration rice varieties, long-duration rice varieties have about 60 days of the vegetative stage. So, in total, the medium duration growth of rice is 115–135 days, assuming farmers apply the transplantation method (Moldenhauer & Slaton, 2001; Yoshida, 1981).

The ratoon method has been known by rice farmers for a long time (Fitri et al., 2019; Juanda, 2016). In the transplantation method, farmers need land preparation and new seed for each new growing season. In contrast, the ratoon method aims to cultivate parent rice for new growing seasons without land preparation and without new seed. Although the ratoon method may save cost and farming time, this method has been neglected due to its low yield compared to its parent yield (Adji, 2022; Fitri et al., 2019; Pasaribu, 2016; Pasaribu & Anas, 2018).

The salibu method aims to improve the ratoon methods as it can sustain or increase rice yield compared to its parent yield. The salibu method is coined by other scholars (Erdiman & Misran, 2013) when they supervised farmers in North Sumatera. After successful harvesting seasons, the salibu method has been widely applied in other Indonesian provinces (Fitri et al., 2019).

The main difference between the two is in the ratoon method, farmers would not prune parent rice as new rice plants, whereas farmers in the salibu method would prune parent plants as new rice for the next growing seasons. Usually, after 5–15 days of the harvesting seasons, farmers would prune the parent rice (Pasaribu, 2016).

Because pruning the parent rice occurs after the rice has stems, roots, and tillers, salibu method combines the vegetative and reproductive phases. In turn, the total

Table 1 Three stages of rice growth

No	The stages of rice growth	
	The transplantation method	The ratoon and salibu methods
I	Vegetative (from seed germination to panicle initiation, 45–65 days) Rice biomass is converted into roots, tillers, and stems through respiration growth mechanisms	Please bear in mind that growth mechanisms in the transplantation and the salibu methods are similar such as panicle initiation, new root, and new tillers. The main differences are explained in this column
II	Reproductive (panicle initiation to flowering and heading, 35 days) Panicle which emergences in the reproductive stage is a critical because it will affect the number of panicles per farming area and grains per panicle. In the reproductive stage, photosynthesis is supported by the leaf	While the transplantation method needs land preparation, the ratoon and the salibu methods do not require land preparation Pruning parent rice after the harvesting seasons (the ratoon method does not need pruning) happens in the salibu method Since parent rice already has a root, stem, and leaf, rice in the earliest stage of the salibu method has been already in the vegetative and reproductive stages (Pasaribu, 2016; Pasaribu & Anas, 2018)
III	Ripening (flowering and heading to mature grain, 30 days) Rice grain is filled with milky materials into a soft dough. Afterward, panicles and leaves turn yellow with mature and hard grain	In principle, the salibu method has a similar length to the ripening stage (Pasaribu, 2016; Pasaribu & Anas, 2018)

Source Bahri (2017), Moldenhauer and Slaton (2001), and Yoshida (1978, 1981)

growing seasons of rice, after the salibu method, would be less than 90 days (Fitri et al., 2019; Marpaung et al., 2022; Paiman, 2021).

2.3 Data Collection

There are two types of collected data: climatic data and non-climatic data. Climatic data such as temperature and rainfall were collected from the Indonesian Bureau of Meteorological and Geophysics (BMKG). Non-climatic data such as rice yield were collected from the Indonesian Statistics Bureau (BPS).

Collected data are statistically analyzed to estimate relationships between rice yield and climate factors. Rice yield could be affected by climate factors including maximum temperature, minimum temperature, and rainfall (Khairulbahri, 2021; Lobell & Gourdji, 2012; Welch et al., 2010). So, this study applies linear regression models to relate rice yield to temperature, rainfall, and time trends (Lobell & Burke, 2010). High-yield rice varieties are based on technological progress that is encapsulated in improved rice varieties (Bahri, 2017; Gnanamanickam, 2009; Lobell & Burke, 2010). As another study (Khairulbahri, 2021; Lobell & Burke, 2010), this study uses time variables as a representation of the technological progress of improved rice varieties. Equation 1 shows a statistical model that estimates relationships between climate, time variables, and rice yield as follows:

$$\mathrm{Yi} = \mathrm{c}_1 + \alpha_1\mathrm{Ai} + \alpha_2\mathrm{Ai}^2 + \alpha_3\mathrm{Ti} + \alpha_4\mathrm{Ti}^2 + \alpha_5\mathrm{Ta} + \alpha_6\mathrm{Ta}^2 + \alpha_7\mathrm{Ri} + \alpha_8\mathrm{Ri}^2 + \varepsilon_1 \quad (1)$$

Yi = rice yield (tons/ha) in year t;
Ai = time variable as a representation of the technological change of rice (a year since 1993);
Ai^2 = squared time variable as a representation of the technological change of rice;
Ti = minimum temperature in Celsius in year t;
Ta = maximum temperature in Celsius in year t;
Ri = seasonal rainfall in mm/year in year t;
c_1 = constant;
ε_i = error.

For time variables, Ai and Ai^2, it was set as 1 starting since 1993 (in 1993 and 2019, Ai = 1 and Ai = 27 respectively). The statistical software Eviews© version 7 was used to conduct several statistical tests assessing the adequacy of a statistical model rice yield.

The performance of selected statistical models will be obtained and assessed by Ordinary Least Square (OLS) and F-tests, respectively. Owing to this, selected statistical models must conform to normality assumptions (Greene, 2003; Gujarati et al., 2012). Furthermore, reliable statistical models have to be fulfilled relevant assumptions such as homoscedasticity, uncorrelated predictors, and no serial correlations (Greene, 2003; Gujarati et al., 2012; Lobell & Burke, 2010).

To obtain the correct model specification, the general to simple approach is applied as it is a suitable practice in modern analysis (Greene, 2003; Quiroga & Iglesias, 2009). In principle, the general to simple approach asks scientists to collect all possible independent variables and regressed them with the dependent variable step by step to obtain the best statistical model(s). By applying the general to simple approach, we can prevent ourselves from abandoning possible significant independent variables, leading to the best statistical model (Greene, 2003; Quiroga & Iglesias, 2009).

3 Results and Discussion

3.1 Observed Data

Table 2 shows the total rice production in West, Central, and East Java, the three main rice provinces in Indonesia. In general, the three provinces have similar rice production, rice yield, and harvested areas. It is also shown in Table 2 that the total rice production of the three main rice provinces is about 50% of the Indonesian total rice production, showing the importance of the three provinces in Indonesia's rice production.

Another similarity between the three provinces is the dominance of irrigated farming areas (95%) instead of non-irrigated areas (5%) (BPS Jawa Barat, 2021; BPS Jawa Tengah, 2021; BPS Jawa Timur, 2021). As consequence, farmers in the three provinces harvest their rice almost throughout the year. Of climate, West Java is relatively cooler than the other provinces but West Java has higher rainfall than the other provinces.

The highest maximum temperature (33.98 °C) and the highest minimum temperature (24.9 °C) are lower than rice's temperature threshold of maximum temperature (35 °C) and minimum temperature (22 °C), respectively (Table 3). However, the impacts of temperature on rice yield are not always statistically significant as described in the following paragraphs.

As seen in Table 4, for all rice in the three provinces, technological progress such as better rice varieties have an important role in increasing rice yield. This premise is in line with existing studies (Bahri, 2017; Gnanamanickam, 2009; Lobell & Burke, 2010), confirming the importance of technological progress in supporting rice yield. However, the impacts of high temperatures are different in the three Indonesian provinces. The following paragraphs explain the different impacts of high temperatures on rice yield.

3.1.1 West Java

As 95% of total farming areas are irrigated rice areas, it is not surprising that the effects of rainfall are not statistically significant on rice yield in West Java. Likewise, the impacts of minimum temperature are not statistically significant, as the highest observed minimum temperature was 21.7 °C, lower than the threshold of minimum temperature (22 °C) (Peng et al., 2004). In another hand, a time variable, as a representation of technological progress such as better rice varieties, significantly can increase rice yield in West Java ($p < 0.01$). Likewise, maximum temperature has a significant impact on rice yield. Every 1 °C increase in maximum temperature tends to increase rice yield by about 3.387 tons/ha ($p < 0.01$).However, the maximum temperature tends to negatively decrease rice yield once the observed maximum temperature surpasses 33 °C. This is in line with other studies (Bahri, 2017; Jagadish et al., 2010; Khairulbahri, 2021), stating that maximum temperature tends to decrease rice yield by over 33 °C.

3.1.2 East Java

Farmers in East Java have experienced the negative impacts of high minimum temperature. The impacts of minimum temperature are statistically significant as

Table 2 Rice production, rice yield, and harvested areas in West Java, Central Java, and East Java

Province	Rice production (tons)			Rice yield (tons/ha)			Harvested areas (ha)			Fractions of rice production		
	2020	2019	2018	2020	2019	2018	2020	2019	2018	2020 (%)	2019 (%)	2018 (%)
West Java	9,016,772	9,084,957	9,647,359	5.7	5.75	5.7	1,586,889	1,578,836	1,707,254	16	17	16
Central Java	9,489,164	9,655,654	10,499,588	57	5.75	5.8	1,666,931	1,678,479	1,821,983	17	18	18
East Java	9,944,538	9,580,934	10,203,213	5.7	5.63	5.8	1,754,380	1,702,426	1,751,192	18	18	17
Indonesia	54,649,202	54,604,033	59,200,534	5.1	5.11	5.2	10,657,275	10,677,887	11,377,934	52	52	51

Source BPS Jawa Barat (2021), BPS Jawa Tengah (2021), BPS Jawa Timur (2021), and BPS Indonesia (2022)

Table 3 A summary of climatic data (1993–2019). Bracketed values show minimum and maximum values

Provinces	Maximum temperature (°C)	Minimum temperature (°C)	Rainfall (mm)
West Java	28.93 (26.1–31.83)	19.21 (16.2–21.7)	2,241 (1,046–4,453)
Central Java	32.28 (31.4–33.98)	24.17 (23–24.9)	3,227 (3,140–3,398)
East Java	33.43 (32–34.35)	22.48 (20.7–24.68)	2,025 (1,218–3,464)

Table 4 Statistical models for each province

Variables	West Java	East Java	Central Java
Time variable	0.052**	0.037**	0.022**
Minimum temperature	–	−0.09*	0.149*
Maximum temperature	3.387**	5.78*	0.118*
Squared Maximum temperature	−0.058**	−0.087*	–
C	−44.38*	−88.7*	−2.36
Adjusted R^2	79%	63%	77%
F-tests	<0.01	<0.01	<0.01

* = significant at $\alpha = 5\%$, ** = significant at $\alpha = 1\%$

the observed minimum temperature is higher than 22 °C, the threshold of minimum temperature (Peng et al., 2004). Table 4 shows that rice yield will be decreased by about 0.09 tons/ha for a 1 °C increase in minimum temperature ($p < 0.05$).

In contrast, farmers have experienced the positive impacts of maximum temperature as 1 °C increase in maximum temperature tends to increase rice yield by about 5.87 tons/ha ($p < 0.05$). However, similar to rice yield in West Java, the maximum temperature tends to decrease rice yield once the maximum temperature surpasses about 33 °C. Rainfall, conversely, has no significant impact as 95% of the total rice farming is supported by irrigation facilities (BPS Jawa Timur, 2021).

3.1.3 Central Java

Surprisingly, observed minimum temperature and observed maximum temperature tend to increase rice yield in Central Java. For every 1 °C increase in the minimum and maximum temperature, rice yield is projected to increase by about 0.149 and 0.118 tons/ha, respectively ($p < 0.05$). The impacts of rainfall on rice yield are not statistically significant ($p > 0.1$), due to a large fraction of irrigated rice land (95%) in Central Java (BPS Jawa Tengah, 1993).

The positive impacts of high temperatures are due to important reasons. The main reason is rice farmers in Central Java have applied the salibu method (Nugroho et al., 2018; Pemerintah Jateng, 2021; Sulistyono, 2019). The salibu method is local wisdom that shortens the growth duration of rice significantly without sacrificing rice

Table 5 Comparisons between the salibu and non-salibu methods

No	Descriptions	Salibu	Non-salibu (ratoon)
1	Fertilizer	Organic	Organic/non-organic
2	Seed	No new seed each season	New seed each season
3	Land cultivation	No	Yes
4	Growth duration	<90 days	>90 days
5	Yield	Relatively higher	Relatively lower
6	Farming cost	Relatively lower	Relatively higher
7	Water consumption	Relatively lower	Relatively higher

Source Budiant (2021), DodiCandra (2020), Fitri et al. (2019), Krisnaputri (2020), Marpaung et al. (2022), Paiman (2021), and Yamaoka et al. (2017)

yield (Jahari & Sinaga, 2019; Nugroho et al., 2018; Wahyuni et al., 2019). The salibu method can shorten rice duration growth, to less than 90 days, enabling farmers to harvest their rice for about 2–3 months (Agustina et al., 2021; Muzabi, 2021). Short growth-duration (SDG) rice varieties after the salibu method also enable rice to escape from climate extremes such as heat stress and droughts (Abdullah et al., 2008; Bouman, 2007; Yoshida, 1981). SDG rice varieties also enable farmers to escape heat stress during the night through transpiration cooling as most farming areas have sufficient irrigated water during the night (Jagadish et al., 2010; Wassmann et al., 2009).

Table 5 highlights comparisons between the salibu and non-salibu (ratoon) methods. Rice farmers who apply the salibu method tend to enjoy some benefits such as fewer farming costs (less labor cost and less seed cost), shorter duration growth, and relatively higher yield. Two advantages of the salibu method such as shorter growth duration and less water consumption are important in coping with the negative impacts of climate change (Lobell & Burke, 2010; Rosenzweig et al., 2020).

Moreover, in the salibu method, rice farmers prune parentrice which leads to new shoots or tillers, leading to new roots (Fitri et al., 2019; Isnawan et al., 2022; Pasaribu & Anas, 2018). More roots and more tillers mean that rice has a higher coverage to capture important nutrients and higher rice yields, respectively. As parent rice is still in the farming field, parent rice and new fertilizer can be a better combination of nutrient sources for rice growth.

Differing farmers in Central Java, several studies showed that farmers in East Java (Afiani, 2018; Sayaka & Hidayat, 2015) and West Java (Dianawati & Sujitno, 2015; Rasmikayati et al., 2020; Rochdiani et al., 2017; Rohaeni & Ishaq, 2015) have sown rice varieties with relatively long growth duration (115–125 days) with limited farmers having applied the salibu method in East (Budianti, 2021) and West Java (Effendy et al., 2021). These are possible explanations for why farmers in West Java and East Java have experienced the negative impacts of high temperatures.

In tackling the negative impacts of high temperatures, rice scientists have developed heat-tolerant rice varieties (Yang et al., 2017; Ye et al., 2015; Zhang et al., 2013).

However, the application of heat-tolerant rice varieties may be lagged due to costs, different rice tastes, and policymakers' involvement (Kondamudi et al., 2012). As shown in this study, sowing SDG varieties after the salibu method should be seen as an alternative to cope with the negative impacts of high temperatures.

Existing studies (Boonwichai et al., 2019; Lobell & Burke, 2010) suggested that shifting crop seasons is one solution for minimizing the negative impacts of high temperatures. The salibu method, due to its shorter-duration growth, enables farmers to apply shifting crop seasons, preventing the rice from heat stress.

Likewise, since the rising temperature is associated with climate change, the positive impacts of the salibu method should be investigated further in higher temperatures, anticipating the negative impacts of higher temperatures beyond the observed temperatures.

4 Conclusion

This study discusses the impacts of climate factors on rice yield in the three Indonesian main provinces including West Java, Central Java, and East Java. This is important as recent temperatures in these main rice producers have reached the temperature threshold for rice growth. The importance of this study is also supported as these three regions capture about half of the Indonesian total rice production. This means that any possible threat to rice production in given provinces is likely to influence the Indonesian rice supply significantly.

In near future, the maximum temperature tends negatively affect rice yield in West Java and East Java if the maximum temperature is higher than 33 °C. Similarly, as the observed minimum temperature has surpassed the threshold of minimum temperature, farmers in East Java have experienced the negative impacts of minimum temperature on their rice. Again, since the recent minimum temperature in these three main regions has been close to the threshold of minimum temperature, changing climate is a big threat to the Indonesian rice supply.

By contrast, farmers in Central Java have experienced the positive impacts of minimum temperature and maximum temperature although both types of observed temperatures have surpassed their temperature thresholds. It appears that salibu method enables farmers to escape heat stress during rice growth, shorten duration growth, and accumulate more grain weight. Thus, this study offers a promising finding that the salibu method can minimize the negative impacts of high temperatures associated with climate change.

One of the possible mechanisms to cope with high temperatures associated with climate change is short-duration growth rice which enables rice to escape from climate extremes (Abdullah et al., 2008; Bouman, 2007; Yoshida, 1981). As the salibu method can shorten rice duration growth, rice farmers in Central Java can escape from the negative effects of high temperatures.

Another study explained other benefits of the salibu such as time saving and less production cost (Fitri et al., 2019). Owing to this, stakeholders such as policymakers/the government and rice scientists should disseminate the benefit of the salibu method as a promising farming method to get a higher profit and to cope with the negative impacts of high temperatures associated with climate change.

As precautionary measures, please bear in mind that the climatic benefit of the salibu method stated in this study is valid within observed temperatures. When the temperature rises significantly, for instance, due to climate change, the climatic benefit of the salibu method should be investigated further.

Last but not the least, the projections of the negative impacts of climate change and simulating possible options to minimize the negative impacts of climate change in the three main provinces will be the next avenue. This is important as IPCC (2013) projects that southern Indonesia will experience an increase in maximum temperature and minimum temperature by about 0.5 °C and 3 °C by 2100, respectively, depending on Radiative Concentration Pathways (RCP) scenarios.

Declarations Competing Interest Statement
The author declares no conflict of interest.

Code Availability
Not applicable.

Authors' Contributions Not applicable.

Funding Statements Not applicable.

Availability of Data and Supporting Material Data used in this study are available on request.
Appendix A provides a summary of statistical test results.

Appendix A: Results of Statistical Tests

Statistical tests	West Java	Central Java	East Java
Jarque–Bera test	$p > 1\%$ Accept the null hypothesis: residuals are normally distributed	$p > 1\%$ Accept the null hypothesis: residuals are normally distributed	$p > 1\%$ Accept the null hypothesis: residuals are normally distributed
Breusch-Godfrey Serial Correlation LM Test	$p > 5\%$ Accept the null hypothesis: data are NOT serial correlated	$p > 1\%$ Accept the null hypothesis: data are NOT serial correlated	$p > 1\%$ Accept the null hypothesis: data are NOT serial correlated

(continued)

(continued)

Statistical tests	West Java	Central Java	East Java
Variance Inflation Factors	<5 There is no any multicollinearity among independent variables	<5 There is no any multicollinearity among independent variables	<5 There is no any multicollinearity among independent variables
Heteroskedasticity Test: White	$p > 1\%$ Accept the null hypothesis: independent variables have homogeneous variances (homoscedasticity)	$p > 1\%$ Accept the null hypothesis: independent variables have homogeneous variances (homoscedasticity)	$p > 1\%$ Accept the null hypothesis: independent variables have homogeneous variances (homoscedasticity)
F-tests	$p < 1\%$ The statistical model can represent relationships between predictors and a predictand properly	$p < 1\%$ The statistical model can represent relationships between predictors and a predictand properly	$p < 1\%$ The statistical model can represent relationships between predictors and a predictand properly

Appendix B: The Results of Statistical Tests

West Java

Dependent Variable: RICEYIELD (Quintal)
Method: Least Squares
Date: 08/17/21 Time: 17:38
Sample: 1993 2019
Included observations: 27

Variable	Coefficient	Std. Error	t-Statistic	Prob
TIMEVARIABLE	0.052330	0.006218	8.415719	0.0000
MAXTEMP	3.387128	1.123246	3.015482	0.0062
MAXTEMP*MAXTEMP	−0.058426	0.019429	−3.007206	0.0063
C	−44.37824	16.26699	−2.728116	0.0120
R^2	0.815006	Mean dependent var		5.398741
Adjusted R^2	0.790876	S.D. dependent var		0.462289
S.E. of regression	0.211405	Akaike info criterion		−0.134125
Sum squared resid	1.027921	Schwarz criterion		0.057851
Log likelihood	5.810685	Hannan-Quinn criter		−0.077040
F-statistic	33.77601	Durbin-Watson stat		1.118468
Prob (F-statistic)	0.000000			

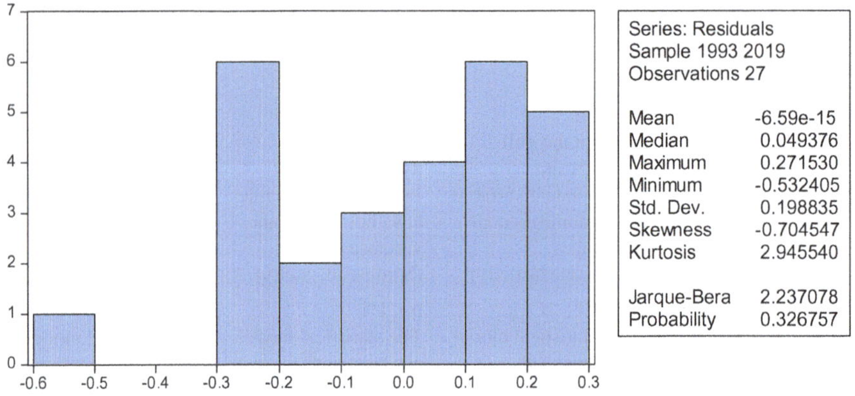

Breusch-Godfrey Serial Correlation LM Test:			
F-statistic	2.282795	Prob. *F*(3,20)	0.1102
Obs*R^2	6.887058	Prob. Chi-Square(3)	0.0756

Variance Inflation Factors
Date: 08/17/21 Time: 17:41
Sample: 1993 2019
Included observations: 27

Variable	Coefficient variance	Uncentered VIF	Centered VIF
TIMEVARIABLE	3.87E-05	5.995509	1.417120
MAXTEMP	1.261682	642,551.5	595.8193
MAXTEMP*MAXTEMP	0.000377	162,660.7	600.2958
C	264.6151	159,862.4	NA

Note Maxtemp and its square are a self-multiplication so their high VIFs are an indication that they have multicollinearity issue(s)

Heteroskedasticity Test: White			
F-statistic	0.715020	Prob. *F*(6,20)	0.6419
Obs*R^2	4.768739	Prob. Chi-Square(6)	0.5738
Scaled explained SS	3.366215	Prob. Chi-Square(6)	0.7617

East Java

Dependent Variable: RICEYIELD
Method: Least Squares
Date: 06/19/21 Time: 22:25
Sample: 1993 2019
Included observations: 27

Variable	Coefficient	Std. Error	*t*-Statistic	Prob
TIMEVARIABLE	0.0371019	0.0066341	5.592580	0.0000
MINTEMP	−0.0909193	0.0527742	−1.722798	0.0990
MAXTEMP	5.777870	3.035728	1.903290	0.0702
MAXTEMP^2	−0.0870791	0.0452201	−1.925673	0.0672
C	−88.74630	51.17422	−1.734200	0.0969
R^2	0.691236	Mean dependent var		55.16333
Adjusted R^2	0.635097	S.D. dependent var		3.627105
S.E. of regression	2.191033	Akaike info criterion		4.572200
Sum squared resid	105.6138	Schwarz criterion		4.812169
Log likelihood	−56.72469	Hannan-Quinn criter		4.643555
F-statistic	12.31294	Durbin-Watson stat		1.545657
Prob(*F*-statistic)	0.000021			

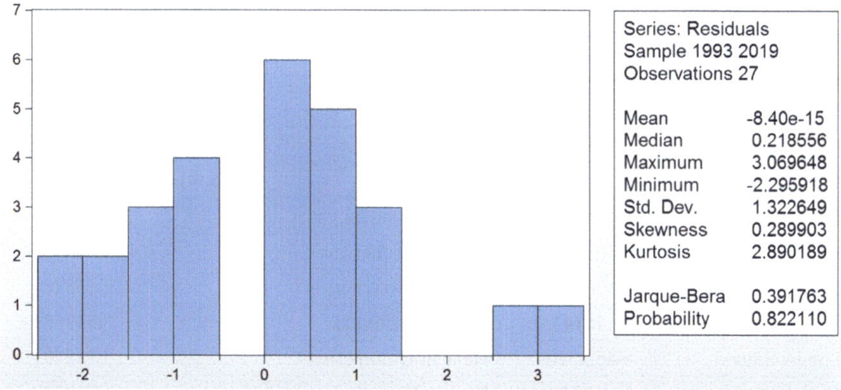

Breusch-Godfrey Serial Correlation LM Test:

F-statistic	1.085246	Prob. *F*(1,21)	0.3094
Obs*R^2	1.326752	Prob. Chi-Square(1)	0.2494

Variance Inflation Factors
Date: 06/19/21 Time: 22:25
Sample: 1993 2019
Included observations: 27

Variable	Coefficient variance	Uncentered VIF	Centered VIF
TIMEVARIABLE	0.004401	6.353340	1.501699
MINTEMP0301	0.278512	783.8395	1.427394
MAXTEMP	921.5646	5,831,862	3391.381
MAXTEMP^2	0.204485	1,459,367	3373.374
C	261,880.0	1,472,883	NA

Heteroskedasticity Test: White

F-statistic	0.892452	Prob. $F(11,15)$	0.5671
Obs*R^2	10.68052	Prob. Chi-Square(11)	0.4704
Scaled explained SS	2.670528	Prob. Chi-Square(11)	0.9944

Central Java

Dependent Variable: RICEYIELD
Method: Least Squares
Date: 06/19/21 Time: 18:04
Sample: 1993 2019
Included observations: 27

Variable	Coefficient	Std. Error	t-Statistic	Prob
C	−2.360660	2.604512	−0.906373	0.3741
MINTEMP	0.1491781	0.0811496	1.838309	0.0790
MAXTEMP	0.1181910	0.0641578	1.842192	0.0784
TIMEVARIABLE	0.0222607	0.0047817	4.655391	0.0001
R^2	0.795775	Mean dependent var		53.83148
Adjusted R^2	0.769137	S.D. dependent var		2.926784
S.E. of regression	1.406266	Akaike info criterion		3.655707
Sum squared resid	45.48444	Schwarz criterion		3.847683
Log likelihood	−45.35204	Hannan-Quinn criter		3.712791
F-statistic	29.87370	Durbin-Watson stat		1.729870
Prob (F-statistic)	0.000000			

Variance Inflation Factors
Date: 06/19/21 Time: 21:22
Sample: 1993 2019
Included observations: 27

Variable	Coefficient variance	Uncentered VIF	Centered
C	678.3484	9261.505	NA
TIMEVARIABLE	0.002286	8.012450	1.893852
MINTEMP0101	0.658526	5287.336	1.432334
MAXTEMP1	0.411622	5857.022	1.643119

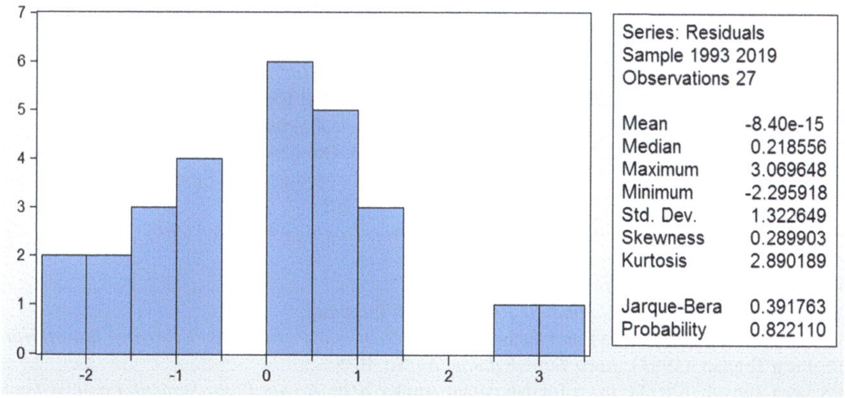

Breusch-Godfrey Serial Correlation LM Test:			
F-statistic	0.162849	Prob. $F(1,22)$	0.6904
Obs*R^2	0.198392	Prob. Chi-Square(1)	0.6560

Heteroskedasticity Test: White			
F-statistic	1.215024	Prob. $F(9,17)$	0.3479
Obs*R^2	10.56913	Prob. Chi-Square(9)	0.3064
Scaled explained SS	7.248406	Prob. Chi-Square(9)	0.6113

References

Abdullah, B., Tjokrowidjojo, S., & Sularjo, S. (2008). *Perkembangan dan prospek perakitan padi tipe baru di Indonesia.*

Adji, R. (2022). Pendampingan Teknologi Budidaya Salibu Pasca Pandemi Covid 19 Di Desa Cimekar, Cileunyi, Bandung. *ACADEMICS IN ACTION Journal of Community Empowerment, 4*(1), 27–39.

Afiani, F. R. K. (2018). Penerapan K-MEANS Clustering Untuk Mengetahui Varietas Padi Unggul Produksi Balai Pengkajian Teknologi Pertanian Jawa Timur. *JATI (jurnal Mahasiswa Teknik Informatika), 2*(1), 336–343.

Agustina, H., Setiawan, B. I., & Sugiyanta, S. (2021). *Manajemen Air Sistem Irigasi Evapotranspirasi Aliran Bawah Permukaan (Sistem Irigasi Evapotranspirasi) Pada Budidaya Padi Sistem of Rice Intensification (SRI) Salibu.*

Amalo, L. F., Hidayat, R., & Sulma, S. (2017). Analysis of agricultural drought in east Java using vegetation health index. *AGRIVITA, Journal of Agricultural Science, 40*(1), 63–73.

Bahri, M. (2017). *Integrating statistical and system dynamics modelling to analyse the impacts of climate change on rice production in West Nusa Tenggara, Indonesia.*

Bandumula, N. (2018). Rice production in Asia: Key to global food security. *Proceedings of the National Academy of Sciences, India Section b: Biological Sciences, 88*(4), 1323–1328.

Boonwichai, S., Shrestha, S., Babel, M. S., Weesakul, S., & Datta, A. (2019). Evaluation of climate change impacts and adaptation strategies on rainfed rice production in Songkhram River Basin, Thailand. *Science of the Total Environment, 652*, 189–201.

Bouman, B. (2007). *Water management in irrigated rice: Coping with water scarcity.* Int. Rice Res. Inst.

BPS. (2020). *Produksi padi Indonesia.* BPS.

BPS Indonesia. (2022). *Tabel Dinamis Produksi Padi Indonesa.*

BPS Jawa Barat. (2021). Jawa Barat dalam Angka 2020. *Badan Pusat Statistik Provinsi Jawa Barat.*

BPS Jawa Tengah. (1993). *Jawa Tengah dalam Angka.* BPS.

BPS Jawa Tengah. (2021). Jawa Tengah dalam Angka 2020. *Badan Pusat Statistik Provinsi Jawa Barat.*

BPS Jawa Timur. (2021). Jawa Timur dalam Angka 2020. *Badan Pusat Statistik Provinsi Jawa Barat.*

Budianti, Y. A. B. Y. A. (2021). Analisis Faktor Produksi Usahatani Padi Dengan Metode Salibu Di Kecamatan Madiun Kabupaten Madiun Jawa Timur. *Jurnal Agribisnis, 23*(2), 274–283.

Chauhan, B. S., Jabran, K., & Mahajan, G. (2017). *Rice production worldwide* (Vol. 247). Springer.

Dianawati, M., & Sujitno, E. (2015, July). *Kajian berbagai varietas unggul terhadap serangan wereng batang cokelat dan produksi padi di lahan sawah kabupaten Garut, Jawa Barat.* In Prosiding Seminar Nasional Masyarakat Biodiversity Indonesia (Vol. 1, No. 14, pp. 868–873).

DodiCandra, Y. T. (Director). (2020). *Tahapan Dan Proses Penerapan Teknologi Padi Salibu Tahap Awal Sampai Tahap Akhir.* https://www.youtube.com/watch?v=cYGszc9rI5Q

Effendy, L., Hanan, A., Haryanto, Y., & Putri, K. (2021). Farmers' preference for innovation of Salibu rice technology in Garut Regency, West Java-Indonesia. *Education, 5*(65), 13.

Erdiman, N., & Misran, M. Y. (2013). Peningkatan produksi padi dengan teknologi spesifik lokasi Sumatera Barat (teknologi salibu). *Laporan Hasil Pengkajian Tahun.*

Falcon, W. P., Naylor, R. L., Smith, W. L., Burke, M. B., & McCullough, E. B. (2004). Using climate models to improve Indonesian food security. *Bulletin of Indonesian Economic Studies, 40*(3), 355–377.

Fitri, R., Kusnadi, N., & Yamaoka, K. (2019). SALIBU technology in Indonesia: An alternative for efficient use of agricultural resources to achieve sustainable food security. *Paddy and Water Environment, 17*(3), 403–410.

Gnanamanickam, S. (2009). Rice and its importance to human life. In *Biological control of rice diseases, progress in biological control* (pp. 1–11). Springer.

Google Maps. (n.d.). *Indonesia and three main Java provinces* [Map]. https://www.google.com/maps

Greene, W. H. (2003). *Econometric analysis*. Pearson Education India.

Gujarati, D. N., Porter, D. C., & Gunasekar, S. (2012). *Basic econometrics*. Tata McGraw-Hill Education.

Isnawan, B. H., Aziez, A. F., & Salisu, M. A. (2022). The role of agronomic factors in Salibu rice cultivation. *The Open Agriculture Journal, 16*(1).

Jagadish, S., Sumfleth, K., Howell, G., Redoña, E., Wassmann, R., & Heuer, S. (2010). Temperature effects on rice: Significance and possible adaptation. *Advanced Technologies of Rice Production for Coping with Climate Change: 'No Regret' Options for Adaptation and Mitigation and Their Potential Uptake* (pp. 19–25).

Jahari, M., & Sinaga, P. H. (2019). Menyiasati penyempitan musim tanam padi dengan budidaya ratun dan salibu. *Dinamika Pertanian, 35*(3), 65–72.

Juanda, B. R. (2016). Potensi Peningkatan Produksi Padi Dengan Meningkatkan IP (Indek Panen) Melalui Penerapan Teknologi Padi Salibu. *Jurnal Penelitian Agrosamudra, 3*(1), 75–81.

Karim, A., Sarra, D., Wasono, R., & Utami, T. (2019). Spatial Modelling for rice production analysis in Central Java Province Indonesia. *Journal of Physics: Conference Series, 1217*(1), 012113.

Khairulbahri, M. (2021). Analyzing the impacts of climate change on rice supply in West Nusa Tenggara, Indonesia. *Heliyon, 7*, e08515.

Kinose, Y., & Masutomi, Y. (2019). Impact assessment of climate change on rice yield using a crop growth model and activities toward adaptation: Targeting three provinces in Indonesia. In *Adaptation to climate change in agriculture* (pp. 67–80). Springer.

Kondamudi, R., Swamy, K. N., Chakravarthy, D. V. N., Vishnuprasanth, V., Rao, Y. V., Rao, P. R., Sarla, N., Subrahmanyam, D., & Voleti, S. R. (2012). Heat stress in rice–physiological mechanisms and adaptation strategies. *Crop stress and its management: Perspectives and strategies* (pp. 193–224). Springer.

Krisnaputri, D. (Director). (2020). *7 Prinsip Teknologi Padi Salibu*. https://www.youtube.com/watch?v=uhZaaEqyYjk

Kuswanto, H., Salamah, M., Retnaningsih, S. M., & Prastyo, D. D. (2018). On the Impact of Climate Change to Agricultural Productivity in East Java. *Journal of Physics: Conference Series, 979*(1), 012092.

Lobell, D. B., & Burke, M. (2010). *Climate change and food security: Adapting agriculture to a warmer world* (Vol. 37). Springer Science & Business Media.

Lobell, D. B., Burke, M. B., Tebaldi, C., Mastrandrea, M. D., Falcon, W. P., & Naylor, R. L. (2008). Prioritizing climate change adaptation needs for food security in 2030. *Science, 319*(5863), 607–610.

Lobell, D. B., & Gourdji, S. M. (2012). The influence of climate change on global crop productivity. *Plant Physiology, 160*(4), 1686–1697.

Lobell, D. B., Schlenker, W., & Costa-Roberts, J. (2011). Climate trends and global crop production since 1980. *Science, 333*(6042), 616–620.

Makbul, Y., Faoziyah, U., Ratnaningtyas, S., & Kombaitan, B. (2019). Infrastructure development and food security in Indonesia: The impact of the trans-java toll road on rice paddy farmer's desire to sell farmland. *Journal of Regional and City Planning, 30*(2), 140–156.

Marpaung, D. S. S., Anika, N., & Bindar, Y. (2022). Strategi Peningkatan Produktivitas Padi Melalui Sistem Salibu Padi. *Jurnal Sumberdaya Lahan, 16*(1), 1–7.

Moldenhauer, K., & Slaton, N. (2001). Rice growth and development. *Rice Production Handbook, 192*, 7–14.

Muzabi, A. K. F. (2021). *Efektivitas Penambahan Pupuk Superfosfat dan Biofertilizer Terhadap Pertumbuhan dan Produksi adi Ratun (Oryza sativa L).*

Naylor, R. L., Battisti, D. S., Vimont, D. J., Falcon, W. P., & Burke, M. B. (2007). Assessing risks of climate variability and climate change for Indonesian rice agriculture. *Proceedings of the National Academy of Sciences, 104*(19), 7752–7757.

Naylor, R. L., & Mastrandrea, M. D. (2009). Coping with climate risks in Indonesian rice agriculture: A policy perspective. In *Uncertainty and environmental decision making* (pp. 127–153). Springer.

Nugroho, A. D., Fadlilah, C. U., Astuti, R. P., Irmania, L. V., Lestari, C., Pinardi, S. T., Anjarwati, N., Anjarwati, A., Bahtiar, E. W., & Pratama, D. A. (2018). Pelaksanaan Program Upaya Khusus (UPSUS) Swasembada Pangan Berkelanjutan di Kabupaten Kendal Provinsi Jawa Tengah. *JPPM (jurnal Pengabdian Dan Pemberdayaan Masyarakat), 2*(2), 287–296.

Paiman, P. Y. (Director). (2021). *Budidaya padi Salibu Part 2.* https://www.youtube.com/watch?v=DhR1Sxo1NtI

Pasaribu, P. O. (2016). *Sifat Fisiologi Dan Agronomi Padi Ratun Dengan Sistem Salibu Pada Budidaya System of Rice Intensification (SRI).*

Pasaribu, P. O., & Anas, I. (2018). Rice ratooning using the salibu system and the system of rice intensification method influenced by physiological traits. *Pertanika Journal of Tropical Agricultural Science, 41*(2), 637–654.

Pemerintah Jateng. (2021). *Petani Padi Purbalingga Terapkan Teknologi 'Salibu Jarwo SuPER'.* https://jatengprov.go.id/beritadaerah/petani-padi-purbalingga-terapkan-teknologi-salibu-jarwo-super/

Peng, S., Huang, J., Sheehy, J. E., Laza, R. C., Visperas, R. M., Zhong, X., Centeno, G. S., Khush, G. S., & Cassman, K. G. (2004). Rice yields decline with higher night temperature from global warming. *Proceedings of the National Academy of Sciences, 101*(27), 9971–9975.

Quiroga, S., & Iglesias, A. (2009). A comparison of the climate risks of cereal, citrus, grapevine and olive production in Spain. *Agricultural Systems, 101*(1–2), 91–100.

Rasmikayati, E., Saefudin, B. R., Rochdiani, D., & Natawidjaja, R. S. (2020). Dinamika Respon Mitigasi Petani Padi di Jawa Barat dalam Menghadapi Dampak Perubahan Iklim serta Kaitannya dengan Pendapatan Usaha Tani. *Jurnal Wilayah dan Lingkungan, 8*(3), 247–260.

Rochdiani, D., Kuswarini, K., & Bobby, R. S. (2017). *Risiko Perubahan Iklim Serta Pengaruhnya Terhadap Pendapatan Petani Usahatani Padi di Jawa Barat.* In Prosiding Seminar Nasional Mitigasi dan Strategi Adaptasi Dampak Perubahan Iklim di Indonesia. Pekanbaru (ID): Universitas Islam Riau.

Rohaeni, W. R., & Ishaq, M. I. (2015). Evaluasi Varietas Padi Sawah pada Display Varietas Unggul Baru (VUB) di Kabupaten Karawang, Jawa Barat. *Agric, 27*(1), 1–7.

Rondhi, M., Fatikhul Khasan, A., Mori, Y., & Kondo, T. (2019). Assessing the role of the perceived impact of climate change on national adaptation policy: The case of rice farming in Indonesia. *Land, 8*(5), 81.

Rosenzweig, C., Mbow, C., Barioni, L. G., Benton, T. G., Herrero, M., Krishnapillai, M., Liwenga, E. T., Pradhan, P., Rivera-Ferre, M. G., & Sapkota, T. (2020). Climate change responses benefit from a global food system approach. *Nature Food, 1*(2), 94–97.

Sayaka, B., & Hidayat, D. (2015). Sistem perbenihan padi dan karakteristik produsen benih padi di Jawa Timur. *Analisis Kebijakan Pertanian, 13*(2), 185–202.

Sembiring, H., A Subekti, N., Nugraha, D., Priatmojo, B., & Stuart, A. M. (2020). Yield gap management under seawater intrusion areas of Indonesia to improve rice productivity and resilience to climate change. *Agriculture, 10*(1), 1.

Serraj, R., & Pingali, P. (2018). *Agriculture & food systems to 2050: Global trends, challenges and opportunities* (Vol. 2). World Scientific.

Setyanto, P., Pramono, A., Adriany, T. A., Susilawati, H. L., Tokida, T., Padre, A. T., & Minamikawa, K. (2018). Alternate wetting and drying reduces methane emission from a rice paddy in Central Java, Indonesia without yield loss. *Soil Science and Plant Nutrition, 64*(1), 23–30.

Sulistyono, A. (2019). *Adaptasi Perubahan Iklim Ala Petani Pulokulon.* http://dispertan.grobogan.go.id/artikel/adaptasi_perubahan_iklim_ala_petani_pulokulon

Sutrisno, J., & Setyono, P. (2017). The effect of adaptation to climate change on rice production in the watershed region of Cemoro, Central Java, Indonesia. *International Journal of Applied Environmental Sciences, 12*(1), 211–220.

Utami, A. W., Cramer, L. A., & Rosenberger, N. (2018). Staple food diversification versus Raskin: Developing climate change resilience in rural Indonesia. *Human Organization, 77*(4), 359–370.

Utami, T., Prahutama, A., Karim, A., & Achmad, A. R. (2019). Modelling rice production in Central Java using semiparametric regression of local polynomial kernel approach. *Journal of Physics: Conference Series, 1217*(1), 012108.

Wahyuni, S., Zulvera, Z., Tanjung, H. B., & Arif, E. (2019). Hubungan Karakteristik Inovasi dan Kearifan Lokal Terhadap Keberlanjutan Penerapan Teknologi Padi Salibu di Kabupaten Tanah Datar, Sumatera Barat. *Jurnal Penyuluhan, 15*(1).

Wassmann, R., Jagadish, S., Sumfleth, K., Pathak, H., Howell, G., Ismail, A., Serraj, R., Redona, E., Singh, R., & Heuer, S. (2009). Regional vulnerability of climate change impacts on Asian rice production and scope for adaptation. *Advances in Agronomy, 102*, 91–133.

Welch, J. R., Vincent, J. R., Auffhammer, M., Moya, P. F., Dobermann, A., & Dawe, D. (2010). Rice yields in tropical/subtropical Asia exhibit large but opposing sensitivities to minimum and maximum temperatures. *Proceedings of the National Academy of Sciences, 107*(33), 14562–14567.

WorldBank, T. (2008). *Adapting to Climate change: The case of rice in Indonesia*. The World Bank.

Yamaoka, K., Htay, K. M., Erdiman, E., & Fitri, R. (2017). *Increasing water productivity through applying tropical perennial rice cropping system (Salibu Technology) In CDZ, Myanmar*.

Yang, Z., Zhang, Z., Zhang, T., Fahad, S., Cui, K., Nie, L., Peng, S., & Huang, J. (2017). The effect of season-long temperature increases on rice cultivars grown in the central and southern regions of China. *Frontiers in Plant Science, 8*, 1908.

Ye, C., Tenorio, F. A., Argayoso, M. A., Laza, M. A., Koh, H.-J., Redoña, E. D., Jagadish, K. S., & Gregorio, G. B. (2015). Identifying and confirming quantitative trait loci associated with heat tolerance at flowering stage in different rice populations. *BMC Genetics, 16*(1), 1–10.

Yoshida, S. (1978). *Tropical climate and its influence on rice [periodicity, productivity and stability]*. (IRRI Research Paper Series [Philippines]).

Yoshida, S. (1981). *Fundamentals of rice crop science*. International Rice Research Institute.

Yuliawan, T., & Handoko, I. (2016). The effect of temperature rise to rice crop yield in Indonesia uses Shierary Rice Model with Geographical Information System (GIS) feature. *Procedia Environmental Sciences, 33*, 214–220.

Zhang, Y., Tang, Q., Peng, S., Zou, Y., Chen, S., Shi, W., Qin, J., & Laza, M. R. C. (2013). Effects of high night temperature on yield and agronomic traits of irrigated rice under field chamber system condition. *Australian Journal of Crop Science, 7*(1), 7–13.

Contribution of Renewable Energy to Social Protection and the Fight Against Climate Change: Case of the Noor 1 Project in Southern Morocco

Amale Laaroussi, Abdelghrani Bouayad, and Ouiame Laaroussi

1 Introduction

Climate change is a source of increased inequality in the world and especially for Southern countries, which are more seriously affected, as well as their growth. Low growth means reduced income, particularly for the first stages of the production chains (small farmers and workers).

This can also affect adaptive capacity. While the rich potentially have a larger capacity to protect themselves against the negative effects of climate change (such as floods, tornadoes, and storms), it becomes harder for the poor to secure sufficient coverage against these risks. Thus, climate change contributes to the perseverance of inequalities, even in developed societies.

Increased inequality could also create social discords in some regions, through water or food resource pressures or spatial conflicts. Climate change has an impact on the most vulnerable as well, such as the aged and young children, who are more susceptible to heat or bad weather. Similarly, during heat waves, homeless people are often exposed to heat and this problem will become more difficult for public health agencies to deal with. In hotter countries or those without sufficient hygiene facilities, the situation is even worse. All the communities living in informal settlements or remote rural zones are expected to struggle to survive in altered climatic conditions. Therefore, climate change has important implications for society: it increases the vulnerability of people who are already exposed.

So climate change has severe consequences for ecosystems. But it is not only an ecological, biodiversity protection, or environmental issue, it also affects human society. Climate change is therefore appearing to be a significant problem for

A. Laaroussi (✉) · A. Bouayad
Faculty of Legal, Economic, and Social Sciences, Moulay Ismail University, Meknes, Morocco
e-mail: laaroussiamale5@gmail.com

O. Laaroussi
Dauphine Research in Management/Paris Dauphine University, Paris, France

N. Shahzad (ed.), *Water and Environment for Sustainability*,
https://doi.org/10.1007/978-3-031-27280-6_7

economic, social, and environmental development. Consequently, the energy transition to renewable energy (RE) has been identified as a central issue for many countries for several years. Switching from fossil fuels to RE is both a huge challenge (technical, administrative, political, and financial obstacles, as well as skills issues) and a great opportunity (economic development, job creation, access to energy, response to climate change) for all countries involved in RE reforms. The energy transition even seems to be a requirement to be able to guarantee greater energy independence while at the same time coping with the growth in energy demand, and it is also a way of responding to the climatic and environmental challenges our planet is facing.

A central concern now is how to match this transition with the sustainable development objectives for countries experiencing rapid change. In the process of energy development, the economic, environmental, and social dimensions necessarily depend on a balance or trade-off in such a process. Thus, the different forms of energy transition can be very different from one continent and country to another. Being aware of the economic, social, and environmental consequences of climate change for Morocco, the country has adopted an energy strategy to face the continuous growth of energy consumption and the problems linked to climate change, which aims at covering 52% of its electricity needs through renewable energy sources by 2030 (Energy, Mines, and the Environment Ministry Report, 2018) while favoring the balance between the environmental, economic, and social dimensions. For Morocco, the transition to RE appears to be an essential option to achieve the targeted objectives.

2 Methodology and Materials

The main steps of the methodology are as follows:

- Literature review on the contribution of renewable energy to social development
- Literature review on the contribution of renewable energy to climate change mitigation
- Vision overview for the transition to RE in Morocco
- Study of the role of the Noor 1 project in social development and the fight against climate change
  - Exposure to the study area
  - Data collection and presentation of analysis tools
  - Analysis of the results

3 Literature Review

3.1 Analysis of the Contribution of Renewable Energies to Climate Change Mitigation

The recent studies by Amponsah et al. (2014) include a detailed review of the literature on the greenhouse gas (GHG) life cycle analysis of RE (79 studies are examined). According to the findings, fossil fuels produce much more greenhouse gas emissions than do renewable energy sources. There is also a consensus that offshore wind technology produces the least amount of greenhouse gas emissions.

The argument about positive effects of using certain about renewable energy (RE) sources in rural Nepali villages was brought up by Sapkota et al. (2014), who made use of the LEAP (Long-range Energy Alternatives Planning) model.[1] The outcome of this model demonstrates that the use of micro-hydro power over the next 20 years will significantly decrease CO_2 emissions by 2553 million tons (Mt). The utilization of solar energy, biogas, and improved stoves will substantially reduce CO2 emissions by 5214 Mt, 35,880 Mt, and 7452 Mt, respectively.

An attempt is made by Shafiei and Salim (2014) to examine the main causes of CO2 emissions based on data from Organisation for Economic Co-operation and Development (OECD) countries (from 1980 to 2011).

In terms of experimental results, the use of renewable energy reduces CO2 emissions, while the use of fossil energy raises them. The recent work conducted by Laaroussi et al. (2021) shows that the transition to renewable energy is a valuable alternative to generating income, reducing CO2 emissions, and promoting sustainable development.

The focus of Creutzig et al. (2014) is on RE in Europe. They assert that the switch to a RE-based energy system can help with both the Eurozone debt problem and climate change. They suggest high levels of policy coordination between Member States in addition to country-specific policy frameworks to improve the transition to RE.

The Intergovernmental Panel on Climate Change (IPCC) (2001) report addressed the consequences of climate change on human well-being. The effects of the climate change process are gradually being felt by people's livelihoods, such as access to water, food production, health, and the environment. If it is not controlled, major economic and ecological disruptions are expected, with disproportionate negative impacts on people and countries in poverty. The impact of climate change, for example, is leading to increased water stress and water shortages and poses a real risk to food security in many countries in Africa, Asia, and Latin America. At the same time, other consequences of climate change on heat wave mortality, vector-borne diseases such as malaria, and access to natural resources will have a direct

---

[1] LEAP Modela software tool for energy policy analysis and climate change mitigation assessment, which had been developed by the Stockholm Environment Institute.

impact on the ability of countries to successfully achieve their poverty eradication and sustainable development goals.

In sum, two major aspects of dealing with climate change are necessary: mitigation and adaptation. Both help to minimize the risks. Mitigation attempts to prevent, or at some point limit, climate change itself, through decreasing greenhouse gas emissions, enhancing energy efficiency, making use of alternative sources such as solar and wind power, and avoiding deforestation. Adaptation, on the other hand, is referred to as deliberate clean energy initiatives aimed at reducing negative effects and exploiting potential opportunities.

3.2 Analysis of the Literature on the Contribution of Renewable Energy Sources to Social Development

Published in February (2011), the authors of the United Nations Environment Programme (UNEP) report affirm that the RE transition is a strategy for a green economy that creates green jobs and is essential for poverty eradication. A coherent economic and social strategy is provided in this report, which aims to spend 2% of the world's GDP on greening the following ten important economic sectors: agriculture, buildings, energy, fisheries, forestry, industry, tourism, transportation, waste, and water.

The Green Economy Initiative is a step toward achieving sustainable development on an international, regional, and national level and not an alternative. The main effects of climate change, persistent poverty, and social exclusion are highlighted in this research along with the green jobs that can be secured as part of the transition to a green economy.. The initiative suggests some strategies to be adopted, such as removing perverse government subsidies, boosting investment in green sectors, etc. According to UNEP, the green economy is the result of the improvement of human welfare and social equity as well as the minimization of environmental risks.

- Job creation

Following Van Jones (2009) "Solar panels do not install themselves. Turbines don't make themselves. Buildings don't waterproof and renovate themselves. Trees in cities, green roofs, and public gardens do not plant themselves. All these activities involve human labor. The acknowledgment of this fact in itself puts into question the myth of ecological restoration always having to be at the opposite end of the spectrum to economic performance".

Pollin et al. (2009) reported that solar energy generates more jobs than fossil fuels. Solar energy generates 5.4 direct jobs per million dollars of production, while coal generates only 1.9 direct jobs and oil and gas only 0.8 direct jobs.

With a focus on the solar energy sector, Wei et al. (2010) found that solar PV employs more people than fossil fuels. Compared to other renewable resources, the PV sector is the most labor-demanding.

In an attempt to determine the quality of jobs created in the RE sector, Sastresa et al. (2010) introduced a quality factor for each RE technology in Aragon (Spain). The outcomes of this study indicate that wind energy can provide better quality jobs than solar thermal and solar photovoltaic.

The United Nations Environment Programme (UNEP) (2008) states: "The scope of green jobs includes a variety of skills, educational backgrounds, and vocational profiles."

The highlight of the quality and variety of jobs created in the renewable energy sector was emphasized by Rifkin, who linked the two twenty-first-century technologies of renewable energy and the internet. This will change how energy is supplied and generate new types of jobs.

In the view of Rifkin (2011), the new workforce of the "third industrial revolution"[2] will require skills in new technical areas, notably the digitized management of electricity networks.

In parallel, numerous studies have explored the potential for the creation of "green" jobs in the European Union (EU). This is mainly motivated by the EU's ambitious goal of ensuring that 20% of its energy production is derived from renewable sources by 2021.

Moreover, there are many methods of measuring jobs created in the renewable energy sector, such as the measurement of jobs per megawatt (MW) installed each year, jobs per MW installed cumulatively, etc. While it is possible to measure the number of jobs created through various ratios, the determination of the quality of the jobs created is a sensitive issue.

It is acknowledged that, in addition to their many obvious advantages in reaching sustainable development, in recent years RE has been recognized as a key element for green post-economic crisis recovery.

4 Vision Overview for the Transition to RE in Morocco

Solar energy had a power generation capacity of 736 MW in Morocco, by the end of 2019. Combined renewables had a generating capacity of 3,264 MW. As solar becomes the cheapest electricity option now, it will take over much of the growth in renewable energy in future. Also, wind energy will be the other main generator of new capacity and energy production. Morocco has a lot of natural wind resources and energy potential. Based in the United Arab Emirates, the Middle East Solar Industry Association (MESIA) (2021) acknowledged in its new report that Morocco is presently a leader in promoting renewable energy. "It noted that Morocco is one of the most advanced countries in the planning and successful implementation of its

[2] The term Third Industrial Revolution (TIR), popularised by Jeremy Rifkin, refers to a new industrial and economic revolution that would be different from the traditional production sectors and would have started at the end of the twentieth century with the development of new information and communication technologies.

renewable energy strategies and that it is targeting the production of 52% renewable energy by the end of 2030 in the capacity mix". Already Morocco is looking at ways to reach 100% renewable energy. This is both impressive and unexpected in itself. In particular, the MESIA report showed how countries around the world could do this in detail, even looking at requirements over short periods throughout every day of the year.

Below is a summary of the vision for the transition to RE in Morocco made by "The Solutions Project" (Fig. 1).

In terms of renewable energy, the Kingdom of Morocco is already very well positioned, when compared to most other countries. Renewable energy creates about

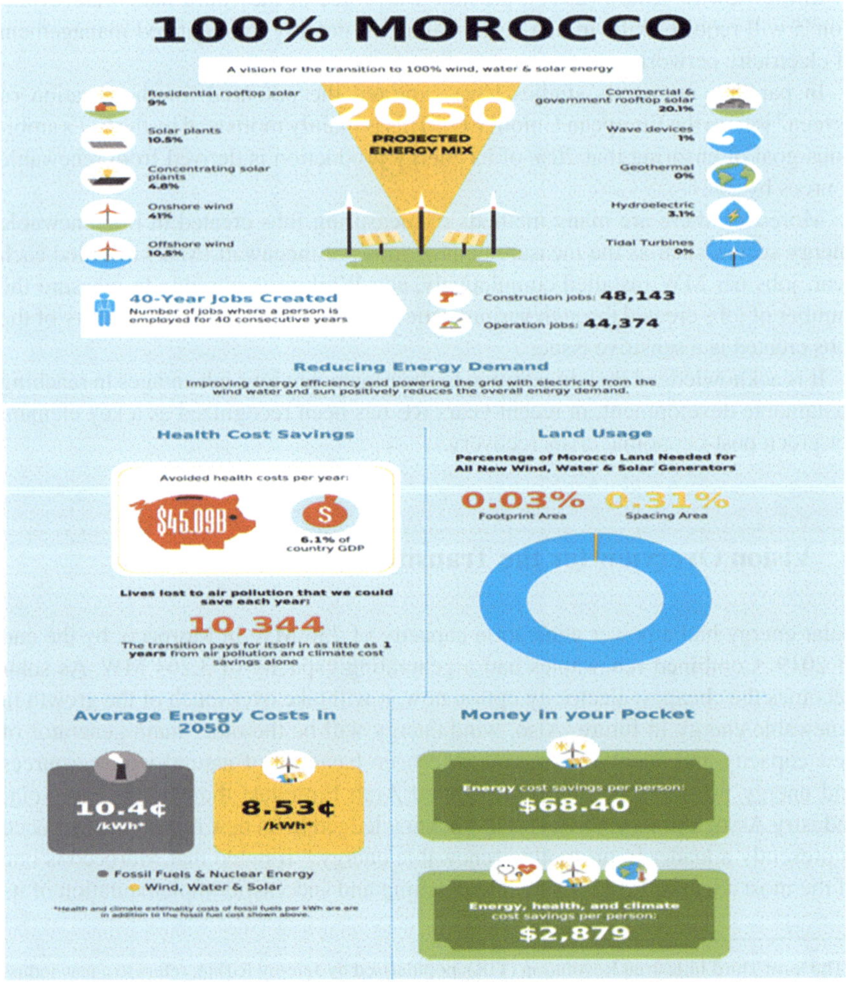

Fig. 1 Vision for the transition to RE in Morocco. *Source* The Solutions Project (2020) Stanford University. https://thesolutionsproject.org/

35% of the country's electricity needs, according to 2019 data. But the country aims to do much better.

By 2030, Morocco intends to be at 52% renewables (for electricity needs), and it aims to reach 100% renewables by 2050.

Morocco's efforts to create green policies have risen quickly since it signed and supported the Paris Agreement. Morocco is fully engaged in the implementation of its renewable energy strategy. Although there are some concerns due to COVID-19, its optimal target of generating 52% of its electricity from renewable sources by 2030 should be reached.

In summary, here are the five lines of action to which Morocco is committed to reaping the full benefits of its three-track strategy to adapt to climate change, mitigate its impact, and create new opportunities:

- By 2030, the country should cover 52% of its electricity needs from renewable energy sources. At the same time, it wants to boost business by setting a target of 35% subcontracting to local agents for the second phase of "NOOR" concentrated solar power plant in southern Morocco.
- The country has removed all support for diesel, petrol, and heavy fuel oil to induce more efficient use of energy and save resources, which will be reinvested in the transition to decarbonized growth.
- The Green Morocco Plan is intended to ensure the protection of the environment but also the livelihoods of all Moroccans. Agriculture, which represents only 15% of GDP, still occupies 40% of the labor force.
- Morocco recognizes that its natural ocean resources are as important as its natural land resources and has therefore improved the management of its coastline and promoted the development of sustainable aquaculture. Fishing accounts for 56% of the country's total exports.
- The country is committed to preserving its groundwater reserves, a natural source of freshwater that is replenished as long as it remains clean and untouched. It is a winning bet for the environment as well as for current and future generations.

In general, it appears that Morocco is making the right and consistent decisions to meet its target of 100% renewable energy by 2050 which will, among other things, prevent deaths from pollution, reduce the costs of global warming, stabilize energy prices, create jobs, and reduce international conflicts over energy.

5 Case Study: NOOR 1 (Concentrated Solar Power) (CSP)-Southern Morocco Region

By implementing the 580 MW solar complex near the city of Ouarzazate, Morocco is considered the first country in North Africa to have developed a stand-alone CSP project. It is the first project to be built under the Moroccan Solar Plan process. The Noor complex—which means "light" in Arabic—was launched by King Mohammed IV in 2013.

The first phase of Noor 1, which has been coordinated by the Moroccan Solar Energy Agency (MASEN), is a 160 MW parabolic mirror power plant with a thermal energy storage capacity and three hours of molten salt wet cooling. Noor 1 consists of the placement of 500,000 parabolic trough mirrors on 800 lines across an estimated 1,308,000 m^2 area.

Steel solar panels are used throughout the project to hold the mirrors and receivers firmly in place. These 393 °C glass panels, which are 12 m high, follow the sun's path to collect and concentrate radiation throughout the course of the day (Fig. 2).

The choice of Noor 1 in southern Morocco (Ouarzazate city) was essentially inspired by the area's high average annual sunshine hours. The region has a significant solar potential, receiving about 2420 $kWh/m^2/year$ of Direct Normal Irradiation (DNI).

- Data collection and analysis techniques and methods

The DPSIR model, which stands for "Driving Forces; Pressure; State; Effect; Response," is used to derive various impact indicators for the qualitative analysis as shown in the Table 1.

This list of indicators helped to identify the different questions asked during the interviews and the literature review, which allows for assessing the social impact of the project on the region where it is located.

Interviewing was the principal source of data collection. A number of semi-structured interviews were carried out with various profiles that were directly or indirectly connected with Noor 1.

A major infrastructure project such as Noor 1 is an opportunity to develop the territory and contribute to the development of the regions in which it is located. Therefore, the implementation of the Noor 1 project in the south of Morocco has been done in compliance with the rules of good governance and international standards in this field. Our study is heavily focused on incorporating local viewpoints by involving the pertinent local stakeholders in the various stages of the process in order to examine the potential positive and negative consequences of the construction and operation phases of the Noor 1 CSP project.

Fig. 2 NOOR 1 (CSP) installation

Table 1 Indicators of social impact included in the above qualitative analysis (Federal Task Force on Sustainable Development [FTSD] and Report on Territorial Dynamic Indicators [RTDI], Table of Sustainable Development Indicators [2002]. http://www.plan.be)

Social impact	Indicator	Objective
	Poverty	– Ensuring quality of life and well-being – Eradication of poverty – Reduction of unemployment to socially acceptable levels – Generalizing the possibility of access to stable and decent employment – Social integration and the fight against illiteracy and school dropout – Increase in life expectancy to socially desirable levels – Quantity and content of public R&D in line with the requirements of sustainable growth
	Employment	
	Human health	
	Working conditions	
	Education, training, and research and development (R&D)	
	Knowledge and technology transfer	
	Skills development	
	Population size and social structure	
	Household wealth and living standards	
	Culture and sense of community	
	Equity and gender equality	
	Participation and social acceptance	

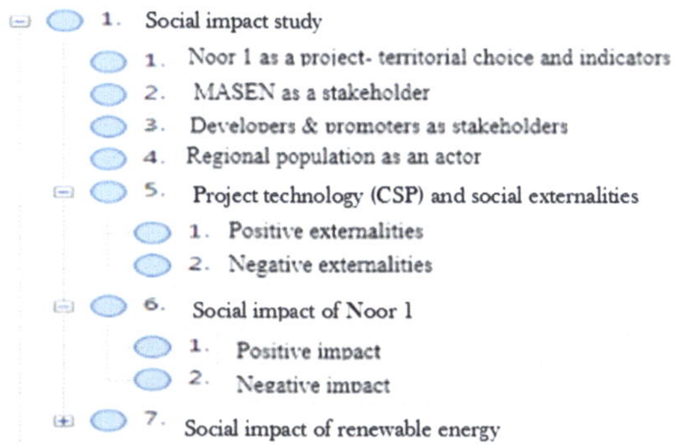

Fig. 3 Node created under Nvivo 12 to handle the verbatims of semi-directive interviews

Collected data are derived from the information gathered through face-to-face conversations (semi-structured interviews) with various respondents who were questioned about the installation of Noor 1 in the Southern region, its mechanism, its technology, and its local and regional social and environmental implications.

The entire coding and analysis strategy is based on the software Nvivo 12 (Fig. 3).

Source coding was carried out to gather all the information collected during the semi-structured interviews and the documentation (Fig. 4).

The data gathered covers every factor (with varying degrees of coverage) that may contribute to the societal impact of Noor 1 on the southern region.

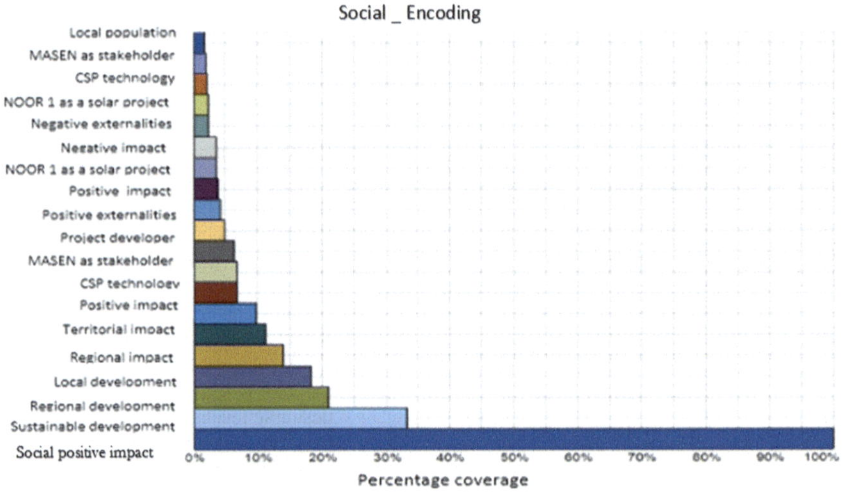

Fig. 4 Encoding by source _ Social dimension

6 Outcomes and Explanations

We used the following grids to categorize and estimate the importance of the social impact of Noor 1 on the southern region.

6.1 Impact Significance

According to the analysis by Nvivo 12, the data collected during the interviews (verbatims) and the documentation, we could rank each impact by its order of importance. On the basis of the Nvivo 12 software's standard default values, five categories of impact importance were established: "Strong," "Substantial," "Moderate," "Trivial," and "no impact" as follows (Fig. 5).

6.2 Status and Type of Impact

The impact status can be divided into two categories: The impact on the chosen indicator is either perceived as being tangible, real, or visible, in which case it is classified as "Perceived impact," or it is predicted using the actual data and outputs, in which case it is classified as "Expected impact."

Furthermore, we have taken into account two types of impact: "positive" represented by the sign (+), and "negative" represented by the sign (−) (Table 2).

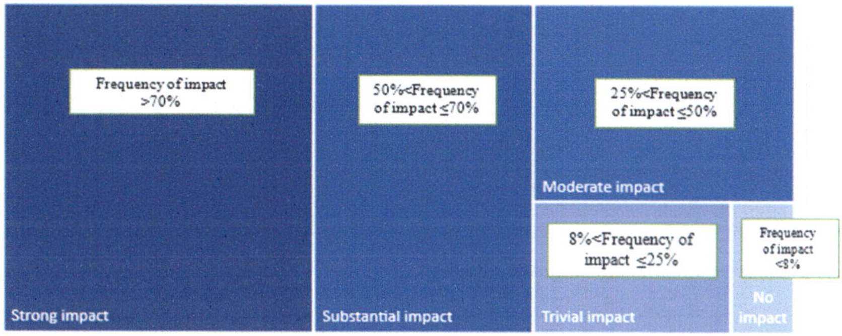

Fig. 5 Impact Significance Reading Grid (Laaroussi et al. 2021). Impact study of NOOR 1 project on the Moroccan territorial economic development. *Renewable Energy and Environmental Sustainability*, 6, 8. Published by EDP Sciences (https://doi.org/10.1051/rees/2021008)

Table 2 Status and type of impact

Status	Type	
Perceived impact	Positive impact	+
Expected impact	Negative impact	–

6.3 *Analysis and Discussion of the Results*

See Table 3.

In the Atlas Mountains of southern Morocco, the Noor 1 project development has taken place in a complex livelihood context, characterized by a combination of difficult sustainability and limited coping capacity, such as:

- Damage to the environment as a result of climate change, soil decomposition, water scarcity, and periodic droughts
- The challenges of education and services, a demographically young population, lack of social infrastructure, social inequalities, and rural–urban migration put social pressure on the region
- Because of high levels of joblessness, rural poverty, and declining agricultural output, economic exclusion is noted

Many factors that could multiply the benefits offered by the implementation of Noor 1 were identified. However, it was clear that the main concerns of the population were employment, education, and health.

As a consequence of Noor 1, new opportunities to create permanent economic activity have arisen. Indeed, the greater availability of energy has facilitated the creation of small and medium-sized enterprises to add to the local companies involved in various maintenance services, guarding, industrial cleaning, etc., which has reduced unemployment and boosted local entrepreneurship. The strengthening of energy capacity has also provided new guarantees and encouragement to investors,

Table 3 Summary of results related to the social impact of Noor 1 project on the southern region of Morocco

Social impact	Status	Type	The initial state in the region	Impact significance	Description
Strengthening family links and social support	Perceived	+	Moderate	Substantial	– There has been a significant improvement in family connections and social support between family members—both in quantity and quality, resulting from the migration of young people who have returned to their families
Poverty eradication and social inclusion	Perceived	+	Trivial	Strong	– Preferential treatment in the supply of labor, goods, and services to local communities in the city of Ouarzazate – Areas have been opened up (70 km of roads) and connected to drinking water and telecoms networks – A special focus on people from the region – Injection of Noor 1's production into the national grid, allowing 600,000 residents to be connected to electricity – National or foreign employees are only called in when local skills are not available – Improvement of the social environment and contribution to the creation of the House of Young Entrepreneurs-800 beneficiaries with over 30 companies created – Improvement of the territorial infrastructure by rebuilding a water supply system for the villages surrounding the power plant
Social conflict, rivals, and envy	Perceived	–	No impact	Moderate	– Social conflicts between community parties and villages, driven by unsatisfied expectations and jealousy toward communities that have benefited from the projects of the social development plan
Unrealizable expectations and frustration	Perceived	–	No impact	Trivial	– Unrealizable expectations and frustrations due to a lack of the understandable and uncertain nature of Noor 1's activities and outcomes among local stakeholders

(continued)

Table 3 (continued)

Social impact	Status	Type	The initial state in the region	Impact significance	Description
Social exclusion and a weakened capacity to engage in decision-making processes	Perceived	–	No impact	Moderate	– The opportunity for local stakeholders to participate meaningfully in the consultation and decision-making process was felt to be limited
The project and its sponsors were viewed with suspicion and the protest of the community	Perceived	–	No impact	Moderate	– Some local stakeholders felt that they were not sufficiently informed and involved, despite plans for job creation and community development, causing unhappiness and resistance to the project, as well as mistrust of the implementing organizations
Psychological well-being and cultural attachment decreased in adjacent communities	Perceived	–	No impact	Trivial	– Due to emotional and cultural attachments, the change of rural landscape to industrial areas has negative consequences for the psychological well-being and sense of alienation of adjacent communities
Decreasing water security in the region	Perceived	–	Trivial	Moderate	– Due to the use of groundwater for building cooling and wet cooling, the accessibility of local water resources has been affected
Living standards and socioeconomic situation of neighboring communities have deteriorated	Perceived	–	No impact	Trivial	– Families who did not have access to work opportunities at Noor 1 are more likely to be exposed to economic disruption due to a reduced capacity to carry out subsistence activities such as goat grazing and firewood collection (particularly for those living in the area around the project)

(continued)

Table 3 (continued)

Social impact	Status	Type	The initial state in the region	Impact significance	Description
Lower living standards for low-income communities	Expected	–	No impact	Trivial	– Students and tourists pushing up prices for local consumers, including those who do not take advantage of the project because of local economic development and demand for products from migrant and domestic workers
Vibration and dust effect on psychological well-being	Perceived	–	No impact	Trivial	– Soil removal and vehicle exhaust emissions of construction or mechanical equipment cause psychological disruption due to increased dust, noise, and vibration – Workers within certain areas of the plant are occasionally subjected to high noise levels, which may have a negative impact on the health of people subjected to such noise levels
Environmental pollution	Expected	–	No impact	Trivial	– Mirror reflections during the operation of Noor 1 and pollution or poisoning of local air, water, and land resources may have likely health impacts
Employment and improvement of working conditions	Perceived	+	Moderate	Substantial	– Net income and employment opportunities through Noor 1 – A boon for tour guides, many of whom are recruited as translators – More advantageous salary levels for the residents of the region – For Noor 1, 1409 workers are hired for the construction project, including 587 workers from the project region (42%) and 35 women – Improving the employability of the local population in the jobs and skills required by Noor 1 – Reserving 70% of the jobs (skilled and unskilled) for the local population

(continued)

Table 3 (continued)

Social impact	Status	Type	The initial state in the region	Impact significance	Description
					– Thousands of jobs were created during the construction phase and hundreds of jobs during the implementation and maintenance phase: 6,000 unskilled jobs in the construction phase and more than 150 jobs in the operational and maintenance phase
Increased awareness of RE systems and society's commitment	Perceived	+	No impact	Substantial	– Enhanced public awareness of RE and the importance of tackling environmental issues, notably climate change, as well as improved commitment and interest in societal responsibility
Improvement of Human Health	Perceived	+	Trivial	Substantial	– Reduction of visual and respiratory diseases due to the use of fuel oil as a source of lighting: Noor 1 provides better access to electricity for people living in rural areas. This will allow the use of fuel oil for lighting to be replaced by electric lamps – Improving access to health services for the population of the region's rural communes – Allowing poor populations to have access to quality care, and bringing medical services closer to poor social categories in rural areas
Improving education, training, and (R&D)	Perceived	+	Trivial	Moderate	– Scholarships are available for students to study abroad – Installation of solar water heaters in schools as well as solar lamps to complement the lighting network, to show children what solar energy is – Reducing school dropouts, particularly among rural girls, by improving school conditions, ensuring access to water and sanitation in schools, and ensuring the safety of students

(continued)

Table 3 (continued)

Social impact	Status	Type	The initial state in the region	Impact significance	Description
					– To stimulate the training and education of technicians and researchers in the field of renewable energy, a new research platform, and professional training programs have been set up – A new institute (Training Institute for RE and Energy Efficiency Professions) is built in Ouarzazate city specifically for RE
Skills development and technology knowledge transfer benefits	Perceived	+	No impact	Moderate	– Many workers have learned how to assemble solar components, which can be used in the ongoing development of other solar projects – Strengthening local capacity, skills development, and R&D efforts enhance the competitiveness and productivity of local industry, generating employment opportunities for local suppliers and workers – Knowledge and skills transfer is guaranteed to support improved agricultural, environmental, economic, and social performance
Household wealth and living standards	Perceived	+	Moderate	Strong	– Through investments in infrastructure and basic community services within the community, the livelihoods of neighboring communities have been improved – The local development plan has made access to important services and livelihoods possible – Provision of low-cost electricity to all households – Supporting students with residence fees at student houses; equipping health centers, sports, and cultural activities

(continued)

Table 3 (continued)

Social impact	Status	Type	The initial state in the region	Impact significance	Description
Growing and strengthening the local feeling of pride and regional reputation	Perceived	+	Moderate	Strong	– Owing to public and media interest, Noor 1 development is seen to be associated with increased local pride and reputation of the southern Moroccan region where the project is sited – Due to the increase in external and foreign migration of workers and students, the rapid changes in the community spirit and cultural identity are affecting the local traditions, standards, attitudes, and lifestyle
Gender equity and equality	Perceived	+	Trivial	Moderate	– Prioritize job creation and economic development for women and youth – Securing the energy supply has enabled women to develop new lucrative activities – Support for women's associations in rural areas

who have not hesitated to relocate to peripheral areas rich in undervalued labor. In addition, the project has facilitated rural and peri-urban electrification programs and provided access to electricity, and reduced the isolation of various communities.

The good quality of electricity provided by The Noor 1 (CSP) project has also improved the production and consumption of goods and services in the southern region through the rise in the price per square meter in the housing market, the enormous activity in the rental market, and the development of infrastructure to secure housing and meals for workers.

Furthermore, as a result of employment and income opportunities, the socioeconomic status and welfare of some households have improved. In concrete terms, Noor 1 has contributed to the improvement of the employability of the population of the trades and skills required for this project, through the creation of thousands of jobs during the construction of the plant and hundreds of jobs during the operation and maintenance phase (several employees have acquired know-how in the assembly of solar components, which they will be able to reuse in the development of other solar projects, as well as the securing of the energy supply has allowed, especially for women, to develop new income-generating activities).

The results also affirm that the intensification of local pride and the gains in regional reputation have proven to have a notable effect on the population, which has benefited from the increase in regional prosperity and value added. Indeed, the reputation of the southern region is enhanced by the project, stimulating new investment and improving health services (such as the reduction of visual and respiratory diseases due to the use of oil as a source of lighting: Noor 1 indirectly allows better access to electricity for rural communities, which will allow the substitution of paraffin for electric lamps for lighting) and education (such as the reduction of school dropout by encouraging students' access to school, by improving school conditions, while ensuring access to water and sanitation in schools and the safety of students). Also, the solar installation has raised awareness of RE, particularly its potential to add value to local economic and social development.

Regarding the negative impacts on social development, the analysis of the results shows that this aspect of the Noor 1 project was of lesser importance and can be presented in two aspects: as increasing the pre-existing sustainability challenges in the region, or as impacts related to the CSP technology or the surrounding context.

One of the project's primary negative aspects is the project's usage of water, which was one of the local stakeholders' top worries. Although numerous surveys indicated that the water demand of Noor 1 CSP will not impair local water supplies, this type of local problem of concern would require a more thorough approach in order to avoid future conflicts.

A further negative impact related to the gap between the quality of training offered by local educational institutions and the specific requirements of the work at Noor 1 was found to be one of the most significant negative impacts, despite efforts to upgrade skills and competencies. Additionally, communities and social groups are not equally represented in the allocation of employment opportunities and social development plan initiatives which negatively affected project ownership, local people's satisfaction, and inter- and intra-community social relations.

Moreover, there is impressive evidence that foreigners, foreign students, and foreign employees have influenced local cultural identity and traditions. The local population's fear is noticed due to the flow of new immigrants adding to the pressure on the limited social services and infrastructure. This concern extends to rising local prices and declining safety and security.

During the construction phase, the increase in vibration, dust, and noise is perceived to be of marginal relevance to stakeholders and residents.

Overall, the implementation of Noor 1 is recognized as an outstanding contributor to regional development and welfare.

Our study shows that RE impacts social and environmental development in a country like Morocco by contributing to:

- Opening up and improving the quality of life of the population. Indeed, RE projects improve living standards and serve as a stimulus for other projects at the local level. For example, access to small amounts of electricity significantly improves the quality of life of users by providing lighting for homes and public places, while reducing pollution of any kind.
- Electrification: Electrification rates in rural and semi-urban zones are boosted by RE projects providing electricity to excluded social groups, thereby increasing the country's power production index.
- Fighting poverty and social exclusion: Access to sustainable energy is now seen as a key factor for sustainable development aimed at poverty reduction. Electric light, modern methods of communication, and access to new media facilitate opportunities for training; cooking; and heating with electricity that is less damaging to health and wildlife and reduces the workload of women and children especially by freeing them from the chores of fetching wood, water, and husking.
- Significant improvement of living conditions in rural areas contribute to limiting rural–urban migration and mitigating the problems associated with accelerated urbanization.
- Improving the skills of poor and disadvantaged people and promoting equality while ensuring the creation of income-generating activities.
- Saving the high cost of repairing environmental damage (mobilizing large financial resources to repair damage and not to develop economic and social activity).
- Avoiding the serious consequences of irreversible damage, constituting a definitive loss (severely contaminated soil, depleted water tables, destroyed species, etc.).
- Safeguarding the living environment of the fauna and flora.
- Providing financial benefits from CO_2 savings.
- Fight against climate change by reducing GHGs of fossil origin and fuel consumption: for example, the GHGs emissions that will be avoided by Noor 1, being 2.9 million tons of CO_2 over 10 years, i.e., 240 thousand CO_2 per year (6 million tons of CO_2 over the 25 years).

7 Conclusion

As a result of this study, we can confirm that the transition to RE is the only means to satisfy the needs of a growing population and promote development, and territorial well-being, reduce GHG emissions, and increase the productivity of natural resources. RE offers governments the opportunity to expand the energy sector and reduce the traditional trade-off between economic growth and environmental protection through technological advances and increasing cost competitiveness.

A unique opportunity for any country is the transition to a RE-based energy system. This transition will balance the adequate energy demand to fuel economic growth and meet the critical requirement to reduce GHG emissions, a major source of climate change.

This is why the United Nations (UN), in its 2030 Agenda for Sustainable Development (2020), has sought to promote government efforts to build the resilience and adaptive capacity of poor and vulnerable people and to promote the integration of climate change measures into national and international policies, strategies, and planning. To allow everyone to benefit from a reasonable basic livelihood, the UN has also advocated for the establishment of social protection systems.

In order to adapt to climate change with social justice, it is necessary to understand how social protection is used to support adaptation to climate change by the world's poorest and most vulnerable households and to reduce poverty.

In the same context, the World Food Programme (WFP) and Oxford Policy Management (OPM) have created a discussion paper which aims to advance knowledge of how social protection might aid in poor and vulnerable households' adaptation to climate change.

For this reason, it suggests some of the climate-related activities that could be combined with social protection programs while drawing on the different theoretical frameworks of analysis that study the links between climate change and social protection and identifies various entry points and conceptions of specific social protection instruments to boost the response to climate change.

Therefore, in Morocco, the deployment of RE technologies has experienced remarkable growth over the past decades, boosted by supportive policies and significant cost reductions. Improving energy security, reducing the adverse effects of climate change, and increasing access to energy are widely seen as reasons for this rise.

Aware of the close interrelation between energy, environment, and sustainable development, Morocco, in its search for sustainable development, should actively engage in the use of sustainable energy resources that do not impact the environment. Therefore, the economic, social, and environmental costs of climate change will be high in the absence of a considerable reduction in GHG emissions. Consequently, it is essential to take action to reduce climate change as much as feasible. To make the transition as smooth as possible, it is necessary to anticipate the strong interactions that mitigation policy has with economic, social, and environmental realities.

References

Amponsah, N. Y., Troldborg, M., et al. (2014). Greenhouse gas emissions from renewable energy sources: A review of lifecycle considerations. *Renewable and Sustainable Energy Reviews, 39*, 461–475.

Creutzig, F., Goldschmidt, J. C., et al. (2014). Catching two european birds with one renewable stone: Mitigating climate change and eurozone crisis by an energy transition. *Renewable and Sustainable Energy Reviews, 38*, 1015–1028.

Federal Task Force on Sustainable Development (FTSD) and Report on Territorial Dynamic Indicators (RTDI), Table of Sustainable Development Indicators. (2002). http://www.plan.be

Intergovernmental Panel on Climate Change (IPCC). (2001). Climate change 2001: Impacts, adaptation, and vulnerability. Contribution of Working Group II to the Third Assessment Report of the Intergovernmental Panel on Climate Change, "Chapter 18: Adaptation in the context of sustainable development and equity" (pp. 877–912). Cambridge University Press.

Jones, V. (2009). Green collar jobs, our planet, the magazine of the United Nations Environment Programme (pp. 8–9).

Laaroussi, A., & Bouayad, A. (2021). Study of economic and environmental impact of photovoltaic installation located in Morocco. Published in 9th International Renewable and Sustainable Energy Conference (IRSEC). IEEE. https://doi.org/10.1109/IRSEC53969.2021.9741203

Laaroussi, A., Bouayad, A., Lissaneddine, Z., & Alaoui, A. (2021). Impact study of NOOR 1 project on the Moroccan territorialeconomic development. *Renewable Energy and Environmental Sustainability*, 6, 8. Published by EDP Sciences. https://doi.org/10.1051/rees/2021008

Middle East Solar Industry Association (MESIA), Solar Outlook Report 2021. https://www.mesia.com/

Pollin, R., Heintz, J., et al. (2009). *The economic benefits of investing in clean energy: How the economic stimulus program and new legislation can boost U.S. economic growth and employment*. Department of Economics and Political Economy Research Institute (University of Massachusetts) and Center for American Progress.

Report of the Ministry of Energy, Mines, and the Environment (MEME). (2018). Morocco. www.mem.gov.ma/

Rifkin, J. (2011). *The Third Industrial Revolution: How lateral power is transforming energy, the economy, and the world* (1st ed., p. 304). Palgrave Macmillan.

Sapkota, A., Lu, Z., et al. (2014). Role of renewable energy technologies in rural communities: Adaptation to climate change in Nepal. *Renewable Energy, 68*, 793–800.

Sastresa, E. L., Uson, A., et al. (2010). Local impact of renewables on employment: Assessment methodology and case study. *Renewable and Sustainable Energy Reviews, 14*(2), 679–690.

Shafiei, S., & Salim, R. A. (2014). Non-renewable and renewable energy consumption and CO_2 emissions in OECD countries: A comparative analysis. *Energy Policy, 66*, 547–556.

The Solutions Project. (2020). Stanford University. https://thesolutionsproject.org/

United Nations. (2020). *The sustainable development goals report*. ISBN 978-92-1-101425-9.

United Nations Environment Program (UNEP). (2008). *Green jobs: Towards decent work in a sustainable low carbon world* (p. 350). Worldwatch Institute.

United Nations Environment Programme (UNEP). (2011). *Towards a green economy: Pathways to sustainable development and poverty eradication. A synthesis for policy maker*. UNEP.

Wei, M., Patadia, S., et al. (2010). Putting renewables and energy efficiency to work: How many jobs can the clean energy industry generate in the US? *Energy Policy, 38*(2), 919–931.

Storage Depletion of Surface Water Reservoirs Due to Sediment Deposition and Possible Management Options

Muhammad Bilal Idrees

1 Introduction

Fresh and drinkable water is a precious resource for life. Every living, breathing species on earth requires water for its survival, and humans are no exception. Water is an essential asset for the existence and survival of humankind. Only 2.5% of the water on the globe is categorized as freshwater, out of which about two-thirds is frozen in glaciers (White, 2012). Merely, one-third of the total freshwater is available for human consumption in the forms of natural reservoirs and streams, e.g., rivers. Rivers have been the primary source of water for humans throughout history, so much so that all ancient human civilizations have originated near riverbanks. Multiple uses of water include irrigation, urban household usage, and various environmental activities. The continuum of all these activities is dependent upon the continued availability of freshwater (Morris & Fan, 1998).

The professionals working in the domain of water engineering are always concerned about the sustainable management of water resources (Kondolf et al., 2014). The two major sources of natural fresh water on earth are surface water and groundwater (Revenga et al., 2000). Surface water, particularly river water, has the greatest potential for sustainable development. Dams must be built to provide reservoir water storage to reliably deliver fresh water from rivers, which is vital for the health and well-being of civilizations. Moreover, reservoir storage is ever more essential as a result of climate change (Cui et al., 2020).

Climate change concerns with rising temperatures, which more or less affect the hydrological processes and other climate variables such as precipitation and evapotranspiration (Gleick, 1989). The frequency of occurrence of hydrological extremes (flood and drought) gets increased as an effect of climate change. The uneven and

M. B. Idrees (✉)
Military College of Engineering (MCE), National University of Sciences and Technology (NUST), Risalpur 23200, Pakistan
e-mail: mbidrees@mce.nust.edu.pk

N. Shahzad (ed.), *Water and Environment for Sustainability*,
https://doi.org/10.1007/978-3-031-27280-6_8

uncertain water availability around the year is problematic for urban water supply and irrigation needs. Keeping in view of the climate change brought about by both natural and anthropogenic factors, water conservation is vitally important, now more than ever (Seitz, 2011).

Retaining structures like dams and weirs have been built to hold back and store floodwater, for subsequent use in drought times. Dams play an important role in the storage and conservation of precious water resources. The dams are built across the river to develop a hindrance in natural river flow. This creates an artificial lake behind the dam body which acts as a reservoir for water storage. This stored water can be used for drinking purposes, and it can fulfill the irrigation requirements as well (Schmutz & Moog, 2018). Moreover, the raised water surface elevation due to reservoir formation increases the navigability of the river (Kim & Lee, 2019).

Dams have numerous advantages, including the ability to store water behind them in the form of reservoirs. Nevertheless, dam construction does come with certain inevitable drawbacks. The main hazard from river water gathering behind the dam body is sedimentation. The reservoir loses valuable water storage capacity as a result of the silt mass that flowing river water carries getting deposited there. Dam safety is impacted by the sedimentation, which makes it difficult to use reservoirs sustainably. Additionally, the reservoir's useful life is decreased by sediment buildup (Schleiss et al., 2016).

The rainfall and subsequent runoff from the watershed erode the topmost layer of the soil. The runoff in the form of hill torrents and small water channels carry the eroded materials and drain into the river. Without dams, soil eroded from the watershed is transported by the rivers into the sea. But when the sediment-laden river flows into the still waters of a reservoir, the drop occurs in water velocities, which reduces its sediment carrying capacity (Mulu & Dwarakish, 2015). The sediment load carried by the river gets deposited in the reservoir, hence reducing the reservoir water storage capacity. The reservoir sedimentation typically follows the deltaic deposition pattern, with the largest sediment load getting deposited near the mouth of the reservoir. The smaller particles may travel further into the reservoir and get deposit near the dam body and the smallest particles remain in suspension. Sedimentation not only reduces the storage capacity of the reservoir; it also hinders its regular operation like power generation. For weirs built for in-channel water storage and to raise the river water elevation, the sedimentation raises the channel bed elevation, which poses flooding risk. Hence, a significant risk factor is contributed by sedimentation toward the overall safety of the dam (Nones, 2019).

Sedimentation affects all reservoirs, lowering their storage capacity. Sedimentation is currently causing a net global loss in storage, which puts domestic water supplies, food and energy production, and flood mitigation at risk (Dominik et al., 2013). China has the highest storage loss rate in the world with lost storage volume accounting for 2.3% of total available reservoir storage annually. The annual storage loss globally due to sedimentation is 0.2–2.5%, with the average value around 1% (Tigrek & Aras, 2011). This translates into the phenomenon that we need to construct 300–400 reservoirs annually just to maintain the current storage value. On average, 33.5 km^3 of the reservoir storage volume is lost every year due to sedimentation

(White, 2012). In some developing countries, storage loss is higher as the construction activities in the watershed areas led to much higher rates of soil erosion.

It is estimated that worldwide there are nearly 25,500 storage reservoirs with a cumulative storage capacity of about 6464 billion cubic meters (BCM). Dams have been constructed all over the world with increased frequency in the last half a century. Hence, most of the desirable dam sites have already been utilized (Schmutz & Moog, 2018). The construction of new dams is also very difficult because of political pressure, the high cost of construction, and land acquisition issues. Moreover, dams also have negative impacts on stream ecology, and it is not an environmentally sustainable solution (Kim et al., 2015). Instead of developing and building new reservoirs, it is vital to maintain the storage capacity of existing ones. By using sediment management and reservoir conservation strategies, it is only possible to reduce the need to construct a new reservoir to the extent possible (Idrees et al., 2019; White, 2012). As a result, sediment management science has become more significant in recent years. By implementing sediment management strategies, water resources engineers must transform the world's inventory of non-sustainable reservoirs into sustainable assets for future generations.

Presently, the storage capacity loss of dam reservoirs due to sedimentation is already higher than the increase of capacity by the construction of new reservoirs for hydropower, irrigation, or domestic use (Dominik et al., 2013) presented a projection of reservoir storage loss for data sets of large reservoirs and subsequent declining water storage capacity. Chinese reservoirs have the shortest life span (22 years) in the world because of high sedimentation rates, while the European and North American reservoirs have a life span of up to 250 years, the highest in the world. The average annual reservoir volume capacity loss because of sedimentation is 1–2%, but a huge decline in storage capacity is observed worldwide (Kondolf et al., 2014). These figures from the literature presented herein highlight the impact of the reservoir sedimentation problem.

It is urgently necessary to consider and assess the reservoir sedimentation issue from a new angle. It is necessary to conduct a detailed and methodical investigation of numerous reservoir sedimentation-related factors from both a technical and environmental perspective. The current study work has made an effort to close the research gap in the body of the literature on reservoir sedimentation. This paper includes a thorough framework for managing and evaluating reservoir sedimentation. Furthermore, the interdisciplinary application of machine learning methods and reservoir conservation modeling has examined reservoir sediment inputs, sediment deposition, and potential reservoir sediment management solutions.

The most recent and significant scientific developments regarding ways to avoid and mitigate sedimentation in reservoirs will be discussed in this study. We'll talk about cutting-edge research techniques for examining sedimentation processes in the context of global climate change. For agencies and departments operating dams and reservoirs, best practices, modeling examples of real-world scenarios, and a check list relating sediment management will be supplied.

2 Mechanism of Reservoir Sedimentation

The loss of storage space caused by sedimentation in dam reservoirs is currently greater than the increase in storage space brought about by the construction of additional reservoirs for hydropower, irrigation, or residential use. Dominik et al. (2013) presented a projection of reservoir storage loss for data sets of large reservoirs and subsequent declining water storage capacity. Chinese reservoirs have the shortest life span (22 years) in the world because of high sedimentation rates, while the European and North American reservoirs have a life span of up to 250 years, the highest in the world. The average annual reservoir volume capacity loss because of sedimentation is 1–2%, but a huge decline in storage capacity is observed worldwide (Kondolf et al., 2014). These statistics highlight the impact of the reservoir sedimentation problem.

Sediment mass is produced by the erosive action of water streams in the watershed and draining toward the river (Samad et al., 2016). Suspended sediment particles are carried by every river, so much so that rivers are treated as bodies of flowing sediments as well as flowing water (Zakaullah et al., 2014). When a river enters the still waters of a reservoir, a decrease in flow velocity and an increase in flow depth and area occurs. The sediment carrying capacity (SCC) is mainly the function of stream velocity; hence, the reduction in SCC occurs (Wang & Hu, 2009). Hence, the sediment mass transported by the river is deposited into the reservoir.

The sediment depositional process starts near the reservoir mouth, with a coarser particle deposit near the mouth of the reservoir. The finer particles cover travels some distance upstream in the reservoir and deposit near the dam body. The finer-most sediment particles remain in suspension due to the upward water current in the reservoir. The segregation and deposition of sediment mass in the reservoir follow the deltaic depositional pattern (Schleiss et al., 2016). Shen (1999) divided the deltaic reservoir sediment deposition into four major parts namely (a) front reach, (b) frontset, (c) topset, and (d) tail reach (Fig. 1). A more detailed sediment deposition pattern for the reservoir was devised by Julien (2010) shown in Fig. 2.

Continuum of reservoir operations would cause smaller sediment particles to migrate closer to the dam body. In the meantime, as more sediment mass would have been brought into the reservoir by the river flows, the depositional pattern may adapt to the shape of a wedge. Hence, the topset of sediment delta is formed with coarser materials. As the smaller particles move further into the reservoir, the bottom set of deposition is composed of finer materials. Fan and Morris (1992b) described the following basic properties of reservoir deltas:

1. The slope abruptly changes between the topset and tail reach deposits.
2. The sediment mass on the topset layer is made up of coarser particles than the sediment mass on the frontset bed. The particle diameter between topset and frontset sediment deposits abruptly changes.
3. The elevation of the zone between the topset and frontset beds is regulated by the reservoir's operational regulations and water surface.

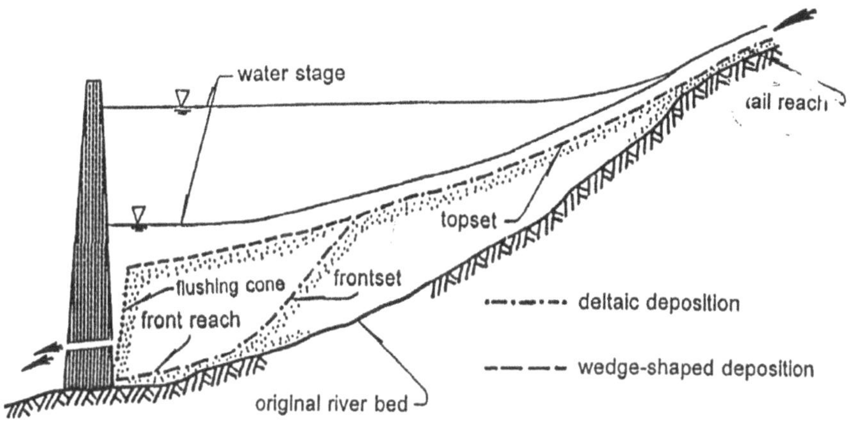

Fig. 1 Conceptual diagram of reservoir sedimentation pattern (Shen, 1999)

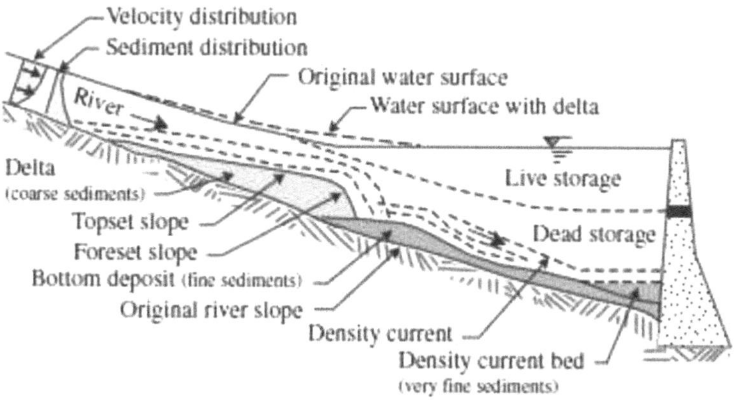

Fig. 2 Detailed reservoir sedimentation pattern (Julien, 2010)

Reservoir sedimentation processes are multiplex and differ depending upon complicated conditions. These conditions range from the entire basin such as watershed sediment yield which in turn depend upon land use and anthropogenic factors. Also, the sediment transport capacity of the river channel, frequency of floods, and cross-sectional and longitudinal geometry of the river plays important roles in reservoir sedimentation. The sediment properties and dam operation rule curves determine the shape and consolidation of the reservoir delta. Morris and Fan (1998) classified longitudinal reservoir sediment depositional patterns into four general types shown in Fig. 3. Typically, coarser fractions of sediment deposits or a larger amount of finer sediment mass like silt shape the deltaic reservoir sedimentation pattern. The tapering and uniform patterns mostly appear in longer reservoirs with fine sediment mass inflow and narrow reservoirs with variable discharge and a small sediment load

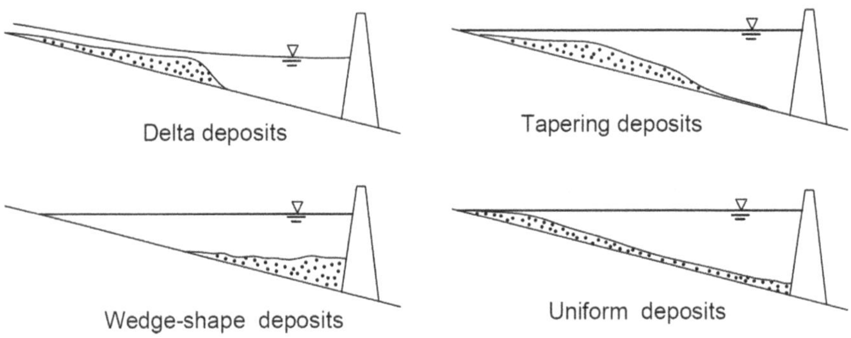

Fig. 3 Classification of reservoir sediment depositional pattern (Morris & Fan, 1998)

inflow, respectively. As the water inflow composed of a large portion of fine sediment mass enters a smaller reservoir, the reservoir sedimentation patterns follow wedge-shaped deposition. The reservoirs with regular water release, the tip of the delta move toward the dam body, and the deltaic deposition transform into a wedge-shaped deposition, which is contemplated as the long-term equilibrium of the reservoirs (Shen & Lai, 1996). The sediment depositional patterns intrinsically are a function of the interactions among water discharge, sediment grain size, channel morphology, and reservoir operations rules.

3 Reservoir Sedimentation Control Measures

The only way to ensure reservoir utilization over the long run is to prevent sediment buildup. Sediment management techniques start in the catchment areas and extend upstream to the river; they are not just used in the reservoir. The ideal combination of answers must be determined by carefully analyzing each circumstance. Figure 4 lists all potential actions and arranges them into categories based on potential application areas (Schleiss et al., 2016). Over the years, several solutions to the problem of reservoir sedimentation have been proposed, but not all of them are practical, efficient, or affordable (Ashraf et al., 2017; Chaudhry, 2014).

It is necessary to control the sediment budget across several reservoirs using an integrated strategy that includes all workable tactics (van Oorschot et al., 2018). An extensive analysis of the sediment issue is part of integrated sediment management, as is the deployment of the sediment methods that are best suited for the specific site. It suggests the importance of operating dams and reservoirs in a way to maintain long-term, sustainable advantages. There are many drawbacks to the existing approach of creating and using reservoirs as non-sustainable assets (Palmieri et al., 2003).

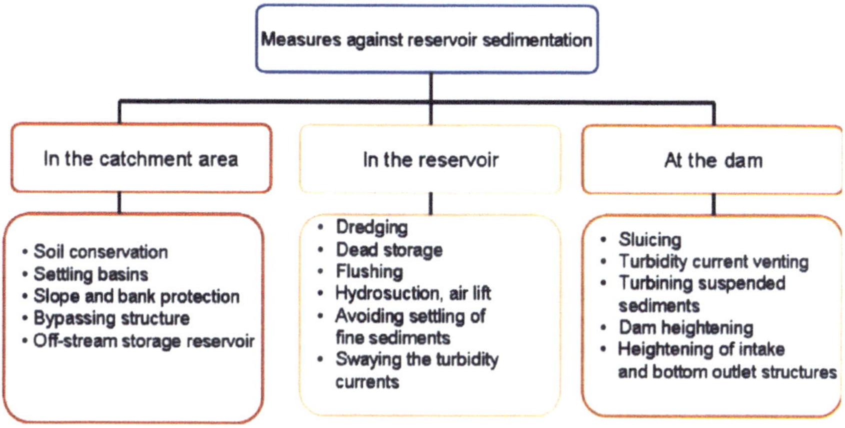

Fig. 4 A list of potential sediment management strategies (Schleiss et al., 2016)

4 Sediment Removal Options in the Reservoir

Methods to manage reservoir sedimentation by the removal of sediments from the reservoir can be classified into two categories: mechanical methods and hydraulic methods. Mechanical methods include Dredging and Trucking. These methods involve the application of heavy machinery and are very expensive. Hydraulic methods incorporate the force of flowing water to remove the sediment deposits (Fan & Morris, 1992a, 1992b) have defined four hydraulic methods to control reservoir sedimentation. These methods are categorized as (1) Sediment routing during flood season, in which floods are passed through reservoirs and hydraulic conditions are maintained such as to minimize sediment deposition in reservoir; (2) Drawdown flushing, in which reservoirs are drawn down to the lowest elevations and previous sediment deposits are exposed and scoured by inflowing water force; (3) Reservoir emptying and flushing, in which reservoir is emptied and previously deposited sediments are scoured by flow over prolonged periods; and (4) Venting the density currents, in which a fraction of sediment inflow is discharged by venting through bottom sluices. Depending upon the particular site conditions, each approach has its level of applicability and associated limitations.

Flushing is scouring deposited sediments and passing the sediment-laden flow through the dam (Atkinson, 1996). Hydraulic flushing is one of the most feasible methods which offer recovering lost reservoir storage without incurring the cost of dredging or other mechanical methods of removing sediment (Emamgholizadeh et al., 2006). Flushing is broadly classified into two categories namely (i) Pressure flushing and (ii) Drawdown flushing. A brief overview of these is given in the following lines.

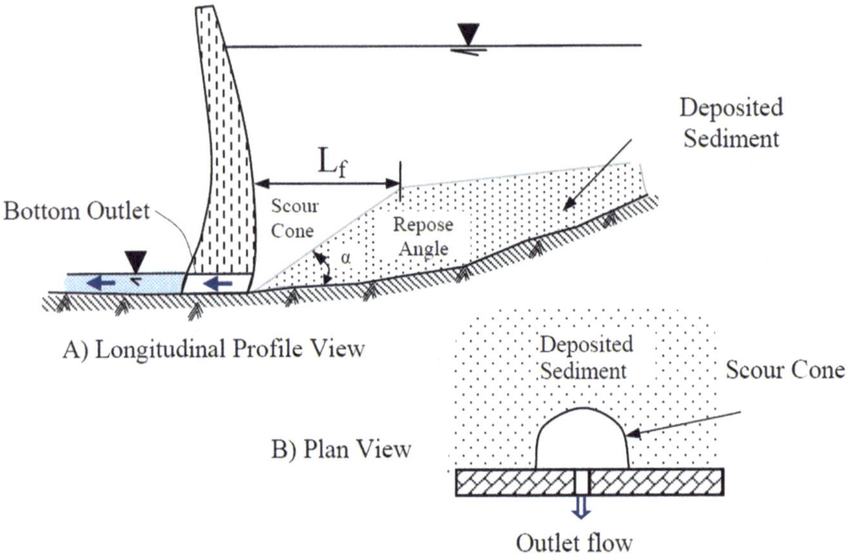

Fig. 5 Plan and longitudinal view of the pressure flushing (Emamgholizadeh et al., 2006)

4.1 Pressure Flushing

Pressure flushing is the process of flushing without lowering the reservoir's water level. Only a small portion of the reservoir is emptied when flushing occurs while a sustained water level is present. A flushing cone forms in the bottom output area, where the flushing is concentrated (Fig. 5).

The water flowing through the hole is clear once the flushing cone has formed and there is no sediment entering the cone. The flushing cone won't lose any sediment because the cone's formation is comparatively stable. Only reservoirs with small reservoir capacities to water influx can use this option (Emamgholizadeh et al., 2006).

4.2 Drawdown Flushing

Drawdown flushing or hydraulic flushing is used for the scouring out of deposited sediments from reservoirs through the use of low-level outlets in a dam by lowering water levels or without lowering water levels. The reservoir level is lowered using the bottom gates in order to increase sediment velocities. This flushing can be classified into under pressure flushing and free-flow flushing. During pressure flushing, water is released through the bottom outlets while a high-water surface elevation is maintained in the reservoir. Free flow flushing means that the reservoir has been emptied and the inflowing water from upstream is routed through the reservoir, re-creating the natural

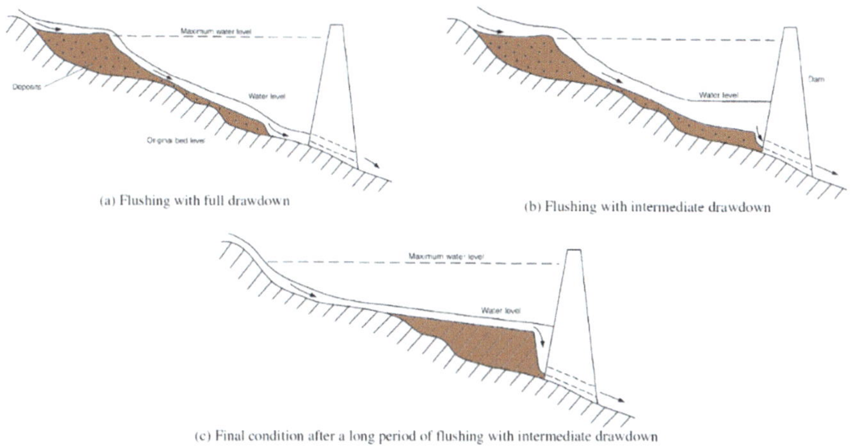

Fig. 6 Longitudinal bed profile of reservoir during flushing (Mahmood & Mundial, 1987)

river flow conditions. Drawdown flushing is efficient as older deposits of sediments are mobilized because of high velocities in the far-off reaches of the reservoir. These deposits are removed or at least they travel further toward the dam body, where they can be removed in future flushing operations.

White et al. (2000) presented three stages of flushing, as shown in Fig. 6. Atkinson (1996) stated that reservoir water level should be drawn down as close to the bed elevation as possible to maximize the flushing efficiency. But partial drawdown can still increase flow velocities at the headwater area where the reservoir delta has formed. In the case of partial drawdown or no drawdown, high flow velocities at the outlets are localized. Drawdown flushing discharges a high concentration of sediments downstream, which may result in severe environmental issues.

Kondolf et al. (2014) stated that it is doubtful whether flushing is effective in larger reservoir. Chaudhry and Rehman (2012) argued that this solution is only suitable for reservoirs with a yearly excess input of water. Apart from this, flushing is found to be highly successful in some reservoirs. These include Baira Reservoir (India), Gebidem Reservoir and Palagneda Reservoir (Switzerland), Gmund Reservoir (Austria), and Hengshan Reservoir (China).

5 Traditional Reservoir Sediment Inflow Estimation Methods

For reservoir sustainability and risk management, a full understanding of fluvial sediment transport processes is essential (Elsey-Quirk et al., 2019; Graf et al., 2010) accurate sediment mass estimation flowing into a reservoir is critical for adopting a suitable sediment management technique. Additionally, for sediment management and the sustainable use of reservoirs, a precise assessment of the sediment influx volume conveyed by the river into the reservoir is required (Annandale et al., 2016). On the basis of field data, observations, and experience, several empirical methods to calculate the volume of reservoir sediment influx have been developed in the past (Garg, 2007). Garg (2007) provided a thorough explanation of these equations. These approaches generally underestimate the sediment inflow volume because of (1) oversimplified assumptions about underlying sedimentation dynamics and (2) not utilizing the reservoir operation parameters as input variables.

The suspended sediment load (SSL) is the major portion (85–90%) of the total sediment load carried by most of the rivers (White, 2012). The conventional technique of sediment rating curve (SRC) has been a widely applied approach for SSL estimation (Moges et al., 2016; Sadeghi & Saeidi, 2010; Zhang et al., 2012). The SRC describes SSL as a function of river water inflow (runoff) and commonly has the power-law form (SSL = aQb). SRC is a simplified method to convert water runoff hydrographs into SSL estimates on very fine temporal resolution (i.e., daily time scale). The sediment inflow data sampling is a very capital-intensive task and hence has financial constraints which make long-term data collection impractical. SRC technique has the advantage that it can be applied to past streamflow records to construct long-term SSL inflow time series. The SRC technique assumes the stationarity of involved hydrological, hydroclimatic, and hydrodynamic variables involved from watershed to the river channel. Hence despite its advantages, SRC has some limitations such as the hysteresis effect. The SRC technique tends to underestimate SSL inflows during high flow seasons (Efthimiou, 2019).

Bathymetric survey (BS) is the technique of computing underwater features which are influenced by navigation, construction, dredging, and related activities in waterways. These surveys are conducted by organizations to determine the existing volume of reservoirs and utilize SONAR and LIDAR. BSs essentially construct the three-dimensional map of the reservoir bottom and using techniques like TIN (triangular irregular network) on GIS software, the existing volume of the reservoir is estimated. For BSs, a variety of techniques can be utilized, including sub-bottom profilers, acoustic doppler current profilers (ADCP), multi-beam surveys, and single beam surveys. The most precise way to calculate the amount of sediment deposited in a reservoir is with BSs (Furnans & Austin, 2008). Such studies are time-consuming, costly, and omit details regarding the origin of the sediment and its affecting elements (Jabbar & Yadav, 2019).

6 Machine Learning Techniques in Reservoir Sedimentation Studies

In contrast, recent studies on reservoir sedimentation and the transport of sediment by rivers have successfully used machine learning (ML) techniques. To establish an explicit connection between SSL and water outflow, Aytek & Kişi (2008) presented a genetic programming approach. Lafdani et al. (2013) applied support vector machine (SVM) and artificial neural networks (ANN) models to forecast daily SSL in the Doiraj River, Iran. Dheeraj et al. (2015) employed an ANN model for rainfall-runoff-sediment modeling with TRMM-3B42 rainfall estimations as input variables. By combining the Dynamic Water Balance Model (DWBM) with ANN, Zhao et al. (2017) assessed the effect of anthropogenic and climatic change factors on sediment load. Khosravi et al. (2018) assessed hourly sediment load influx at the Adean catchment in Chile using stand-alone and hybrid ML models. Malik et al. (2019) applied the gamma test to assess the performance of various ML models for suspended sediment load concentration (SSC) modeling in the Godavari River in India. Huang et al. (2019) used a numerical model in conjunction with machine learning models to estimate half-hourly SSL in Taiwan's Shi-Men reservoir. In order to anticipate outflow sediment concentration for density current venting in reservoirs, Qing et al. (2020) combined ML techniques with the time series analysis.

Streamflow has only been used as an input variable for ML models to predict sediment inflows in these earlier studies (Aytek & Kişi, 2008; Dheeraj et al., 2015; Malik et al., 2019). The logic behind this is to replicate the SRC, which is built using streamflow as an independent variable and SSL as a dependent variable, the only input variable chosen was streamflow (Efthimiou, 2019). Three parameters—streamflow, water temperature, and electric conductivity—were employed by Khosravi et al. (2018) to forecast SSL using three ML models. Idrees et al. (2021) applied and evaluated various ML models based on six (6) machine learning techniques (ANN, ANFIS, RBFNN, SVM, GP, DL) for the suspended sediment load (SSL) inflow prediction from a river to a reservoir.

For SSL projections, none of these studies take reservoir operation characteristics like dam outflow and water stage into account. As a result, there is still a lack of ML model research and application for reservoir SSL investigations. Additionally, there is very little discussion of the findings' global application because they are restricted to a certain geographic area. Therefore, a thorough investigation is needed into the evaluation of common ML models for SSL inflow predictions and how they relate to reservoir operation factors.

7 Conclusions

A comprehensive overview of the recent scientific research in the field of reservoir sedimentation estimation and management has been presented in this study. The complete mechanisms of reservoir sedimentation and all possible control measures were discussed, with sediment flushing operation discussed in detail. Based on this literature review, following conclusions can be drawn:

- Climate change has exacerbated the global rates of soil loss from watersheds and consequently the rates of reservoir sedimentation has increased manifold.
- The annual loss of reservoir storage by sediment deposition is already higher than the storage acquired by building new water reservoirs.
- Possible reservoir sediment management measures range from at watershed scale to at the dam body, and the best management measure should be selected keeping in view the climate change.
- Sediment flushing presents a cheap, viable, and practical measure to recover lost reservoir storage.
- The proper calculation of river suspended sediment load (SSL) and deposition in space and time is essential for the deployment of any sediment management strategy.

The results of this study can serve as suggestions for the execution of operations to flush silt from reservoirs. Additionally, it is anticipated that the observations will be helpful for assessing future dam safety and danger as well as for ensuring reservoir operating sustainability through thorough sediment management.

References

Annandale, G., Morris, G., & Karki, P. (2016). *Extending the life of reservoirs: Sustainable sediment management for dams and run-of-river hydropower, concrete international*. World Bank. https://doi.org/10.1596/978-1-4648-0838-8.

Ashraf, A., et al. (2017). Physically based runoff and sediment modelling of Rawal watershed. *International Research Journal of Engineering and Technology, 4*(12), 543–553.

Atkinson, E. (1996). *The feasibility of flushing sediment from reservoirs* (Report OD 137, TDR Project R5839). Wallingford. http://eprints.hrwallingford.co.uk/111/.

Aytek, A., & Kişi, Ö. (2008). A genetic programming approach to suspended sediment modelling. *Journal of Hydrology, 351*(3), 288–298. https://doi.org/10.1016/j.jhydrol.2007.12.005.

Chaudhry, M. A. (2014). *Numerical simulation for sediment flushing in reservoirs*. University of Engineering and Technology.

Chaudhry, M. A., & Rehman, H. (2012). Worldwide experience of sediment flushing through reservoirs. *Mehran University Research Journal of Engineering & Technology, 31*(3), 254–7821.

Cui, T., et al. (2020). Development of a comprehensive framework for assessing the impacts of climate change and dam construction on flow regimes. *Journal of Hydrology, 590*, 125358.

Dheeraj, K., et al. (2015). Modeling suspended sediment using artificial neural networks and TRMM-3B42 Version 7 rainfall dataset. *Journal of Hydrologic Engineering, 20*(6), C4014007. https://doi.org/10.1061/(ASCE)HE.1943-5584.0001082.

Dominik, W., et al. (2013). Beyond peak reservoir storage? A global estimate of declining water storage capacity in large reservoirs. *Water Resources Research*, 49(9), 5732–5739. https://doi.org/10.1002/wrcr.20452.

Efthimiou, N. (2019). The role of sediment rating curve development methodology on river load modeling. *Environmental Monitoring and Assessment, 191*(2), 108. https://doi.org/10.1007/s10661-018-7167-4.

Elsey-Quirk, T., et al. (2019). Mississippi river sediment diversions and coastal wetland sustainability: Synthesis of responses to freshwater, sediment, and nutrient inputs. *Estuarine, Coastal and Shelf Science, 221*, 170–183. https://doi.org/10.1016/j.ecss.2019.03.002.

Emamgholizadeh, S., Bina, M., & Ghomeyshi, M. (2006). Investigation and evaluation of the pressure flushing through storage reservoirs. *ARPN Journal of Engineering and Applied Sciences, 1*(4), 7–16.

Fan, J., & Morris, G. L. (1992a). Reservoir sedimentation II: Reservoir desiltation and long-term storage capacity. *Journal of Hydraulic Engineering, 118*(3), 370–384.

Fan, J., & Morris, G. L. (1992b). Reservoir sedimentation I: Delta and density current deposits. *Journal of Hydraulic Engineering, 118*(3), 354–369. https://doi.org/10.1061/(ASCE)0733-9429(1992)118:3(354).

Furnans, J., & Austin, B. (2008). Hydrographic survey methods for determining reservoir volume. *Environmental Modelling & Software, 23*(2), 139–146. https://doi.org/10.1016/j.envsoft.2007.05.011.

Garg, S. K. (2007). *Irrigation engineering and hydraulic structures* (21st ed.). Khanna Publishers. https://www.abebooks.com/9788174090478/Irrigation-Engineering-Hydraulic-Structures-S.K-8174090479/plp.

Gleick, P. H. (1989). Climate change, hydrology, and water resources. *Reviews of Geophysics, 27*(3), 329–344. https://doi.org/10.1029/RG027i003p00329.

Graf, W. L., et al. (2010). Sedimentation and sustainability of western American reservoirs. *Water Resources Research, 46*(12). https://doi.org/10.1029/2009WR008836.

Huang, C.-C., et al. (2019). Interdisciplinary application of numerical and machine-learning-based models to predict half-hourly suspended sediment concentrations during typhoons. *Journal of Hydrology, 573*, 661–675. https://doi.org/10.1016/j.jhydrol.2019.04.001.

Idrees, M. B., Lee, J.-Y., & Kim, T.-W. (2019). Estimating RESCON model parameters for efficient sediment flushing in a dam reservoir. *Environmental Earth Sciences, 78*(6), 1–10. https://doi.org/10.1007/s12665-019-8229-9.

Idrees, M. B., et al. (2021). Comprehensive evaluation of machine learning models for suspended sediment load inflow prediction in a reservoir. *Stochastic Environmental Research and Risk Assessment* [Preprint]. https://doi.org/10.1007/s00477-021-01982-6.

Jabbar, Y. C., & Yadav, S. M. (2019). Development of a relationship between hydrometric and hydrographic observations to predict reservoir capacity loss. *Acta Geophysica, 67*(5), 1451–1469. https://doi.org/10.1007/s11600-019-00323-0.

Julien, P. Y. (2010). *Erosion and sedimentation*. Cambridge University Press.

Khosravi, K., et al. (2018). Quantifying hourly suspended sediment load using data mining models: Case study of a glacierized Andean catchment in Chile. *Journal of Hydrology, 567*, 165–179. https://doi.org/10.1016/j.jhydrol.2018.10.015.

Kim, J.-S., et al. (2015). A case study of regional risk assessment of river restoration projects: Nakdong River Basin, South Korea. *Journal of Water and Climate Change, 6*(3), 628–637. https://doi.org/10.2166/wcc.2015.113.

Kim, K.-M., & Lee, J.-W. (2019). Prediction of topographic change of the Nakdong River estuary and analysis of contribution according to external force conditions. *Journal of the Korean Institute of Navigation and Port Research, 43*(1), 64–71. https://doi.org/10.5394/KINPR.2019.43.1.64.

Kondolf, G. M., et al. (2014). Sustainable sediment management in reservoirs and regulated rivers: Experiences from five continents. *Earth's Future, 2*(5), 256–280. https://doi.org/10.1002/2013EF000184.

Lafdani, E. K., Moghaddam Nia, A., & Ahmadi, A. (2013). Daily suspended sediment load prediction using artificial neural networks and support vector machines. *Journal of Hydrology, 478*, 50–62. https://doi.org/10.1016/j.jhydrol.2012.11.048.

Mahmood, K., & Mundial, B. (1987) Reservoir sedimentation: impact, extent, and mitigation. Technical paper (No. PB-88-113964/XAB; WORLD-BANK-TP-71). International Bank for Reconstruction and Development, Washington, DC (USA).

Malik, A., et al. (2019). Evaluating the performance of four different heuristic approaches with Gamma test for daily suspended sediment concentration modeling. *Environmental Science and Pollution Research, 26*(22), 22670–22687. https://doi.org/10.1007/s11356-019-05553-9.

Moges, M. A., et al. (2016). Sediment concentration rating curves for a monsoonal climate: Upper Blue Nile. *The Soil, 2*(3), 337.

Morris, G. L., & Fan, J. (1998) *Reservoir sedimentation handbook: Design and management of dams, reservoirs, and watersheds for sustainable use.* McGraw-Hill. https://searchworks.stanford.edu/view/3757391.

Mulu, A., & Dwarakish, G. S. (2015). Different approach for using trap efficiency for estimation of reservoir sedimentation. An overview. *Aquatic Procedia, 4*(Icwrcoe), 847–852. https://doi.org/10.1016/j.aqpro.2015.02.106.

Nones, M. (2019). Dealing with sediment transport in flood risk management. *Acta Geophysica, 67*(2), 677–685. https://doi.org/10.1007/s11600-019-00273-7.

Palmieri, A., et al. (2003). *Reservoir conservation volume I. The RESCON approach.* Internation Bank for Reconstruction and Development/The World Bank.

Qing, Y., et al. (2020). Ultra-high resolution regional climate projections for assessing changes in hydrological extremes and underlying uncertainties. *Climate Dynamics, 55*(7), 2031–2051.

Revenga, C., et al. (2000) *Freshwater systems.* World Resources Institute.

Sadeghi, S. H., & Saeidi, P. (2010). Reliability of sediment rating curves for a deciduous forest watershed in Iran. *Hydrological Sciences Journal, 55*(5), 821–831.

Samad, N., et al. (2016). Sediment yield assessment and identification of check dam sites for Rawal Dam catchment. *Arabian Journal of Geosciences, 9*(6), 466. https://doi.org/10.1007/s12517-016-2484-9.

Schleiss, A. J., et al. (2016). Reservoir sedimentation. *Journal of Hydraulic Research, 54*(6), 595–614. https://doi.org/10.1080/00221686.2016.1225320.

Schmutz, S., & Moog, O. (2018). Dams: Ecological impacts and management. In S. Schmutz & J. Sendzimir (Eds.), *Riverine ecosystem management: Science for governing towards a sustainable future* (pp. 111–127). Springer International Publishing. https://doi.org/10.1007/978-3-319-73250-3_6.

Seitz, R. (2011). Bright water: Hydrosols, water conservation and climate change. *Climatic Change, 105*(3), 365–381.

Shen, H. W. (1999). Flushing sediment through reservoirs. *Journal of Hydraulic Research, 37*(6), 743–757. https://doi.org/10.1080/00221689909498509.

Shen, H. W., & Lai, J. S. (1996). Sustain reservoir useful life by flushing sediment. *International Journal of Sediment Research, 11*(3), 10–17.

Tigrek, S., & Aras, T. (2011). *Reservoir sediment management.* CRC Press.

van Oorschot, M., et al. (2018). Combined effects of climate change and dam construction on riverine ecosystems. *Ecological Engineering, 120*, 329–344. https://doi.org/10.1016/j.ecoleng.2018.05.037.

Wang, Z., & Hu, C. (2009). Strategies for managing reservoir sedimentation. *International Journal of Sediment Research, 24*(4), 369–384. https://doi.org/10.1016/S1001-6279(10)60011-X.

White, W. R., et al. (2000). *Guidelines for the flushing of sediment from reservoirs.* Wallingford.

White, W. R. (2012). Sediments, flushing from reservoirs. In L. Bengtsson, R. W. Herschy, & R. W. Fairbridge (Eds.), *Encyclopedia of lakes and reservoirs* (pp. 710–716). Springer Netherlands. https://doi.org/10.1007/978-1-4020-4410-6_159.

Zakaullah, B., et al. (2014). Appraisal of sediment load in rainfed areas of Pothwar region in Pakistan. *Global Journal of Researches in Engineering, 14*(6), 25–33. https://globaljournals.org/item/4415-appraisal-of-sediment-load-in-rainfed-areas-of-pothwar-region-in-pakistan.

Zhang, W., et al. (2012). Estimating suspended sediment loads in the Pearl River Delta region using sediment rating curves. *Continental Shelf Research, 38*, 35–46.

Zhao, H., et al. (2017). Quantifying anthropogenic and climatic impacts on sediment load in the sediment-rich region of the Chinese Loess Plateau by coupling a hydrological model and ANN. *Stochastic Environmental Research and Risk Assessment, 31*(8), 2057–2073. https://doi.org/10.1007/s00477-017-1381-4.

Exploring Synergies Between Social Protection Programs, DRR and Climate Change Policies in Pakistan: Case Study NDRMF

Ehsan Ullah and Naeem Shahzad

1 Introduction

Natural hazards trigger natural disasters like earthquakes, hurricanes and floods, etc., and these natural disasters have a devastating effect on social and economic aspects. Human activities, e.g., deforestation, overgrazing, and poor farming, can be one reason for these disasters; secondly, hydrometeorological hazards (floods) and geophysical hazards (earthquakes, landsliding) can also result in disasters (Vakis, 2006). Response to these disasters differs depending on the speed of the disaster like slow onset disaster or rapid disaster. The scale of disasters like localized disasters or countrywide disasters has different consequences which needs to be considered while devising policies to respond to or mitigate disasters.

The focus of **disaster management** remained on disaster response and relief efforts up to the year 1990. The period from 1990 to 1999 was declared as the International Decade for Disaster Risk Reduction **(DRR)** to protect against hazards due to the increased number of extreme events over the last two decades. International Decade for Disaster Risk Reduction (IDNDR) was replaced by United Nations International Strategy for Disaster Risk Reduction (UNISDR) to proceed from protection against hazards to the management of risk through the integration of risk reduction into sustainable development. A paradigm shift from disaster response to a holistic and proactive process of disaster risk management can be seen in recent DRM frameworks (Khan et al., 2022). The focus of **DRR policies** and interventions has also shifted far beyond relief/response activities. Now mitigation and prevention actions have prevalent weightage, especially after major disasters like the earthquakes of Iran (2003), Haiti (2010), and Pakistan (2005), and Indian Ocean tsunami (2004).

This shift in focus is evident from the Hyogo framework of action and the Sendai Framework. Hyogo framework of action for DRR was adopted in the year 2005 by

E. Ullah · N. Shahzad (✉)
National University of Sciences and Technology, Islamabad, Pakistan
e-mail: naeemshahzad@mce.nust.edu.pk

N. Shahzad (ed.), *Water and Environment for Sustainability*,
https://doi.org/10.1007/978-3-031-27280-6_9

the comity of world nations under the umbrella of the United Nations (UN) for the reduction of disaster losses.

Another aspect of looking at this paradigm shift toward a proactive approach is building climate resilient societies by enhancing their coping capacities against environmental hazards (Akter & Mallick, 2013). **Climate change** is the biggest reality of the present day. Climate change is not only a hazard for certain climate-related disasters but also has a severe impact on the communities to increase their social, and economic vulnerabilities. United Nations (UN) has taken preventive and mitigation measures against climate change through different frameworks and agreements like Framework Convention on Climate Change (UNFCCC), Paris Agreement, and Kyoto Protocol. A lot of research and attention is being paid to Climate Change Adaptation (CCA) due to prevalent climate threats (Smit & Pilifosova, 2003) and (Engle et al., 2014).

Disasters result in wider scale loss and damage to physical, economic, social, and environmental infrastructure. However, this damage/loss is disproportionate having a greater impact on the poor and marginalized by further increasing disparities between the haves and have-nots. Thus, poverty is a core vulnerability (Tandon & Hasan, 2005). The resilience of a community is inversely linked with poverty. The poor and marginalized households will have disproportionate physical, social, and economic impact of a disaster (Akter & Mallick, 2013). Poor and marginalized groups are exposed to vulnerabilities especially in the developing world due to climate change and other global processes/hazards. UN designed Millennium Development Goals (MDGs) and Sustainable Development Goals (SDGs) to cover this aspect of poverty alleviation and climate-related issues.

Poverty alleviation improves the financial condition, reduces vulnerability, and increases coping capacity. Governments try to reduce vulnerabilities through the implementation of Social Protection programs (SP), Climate Change Adaptation (CCA), and Disaster Risk Reduction (DRR) policy frameworks. SP, DRR, and CCA are three streams working almost for the same cause; however, most of the interventions are operational in isolation in their respective domains (Heltberg et al., 2009).

Harmony among SP, DRR, and CCA is necessary to bring efficiency and effectiveness in the utilization of human material and financial resources which are always limited in nature. Resource constraints become more evident in the event of disasters. Pakistan is a developing country and is no exception as far as resources for poverty alleviation, climate change adaptation measures, and disaster management are concerned. So the question arises of how to bring integration among these streams in Pakistan. The research aims to infer the role of adaptive social protection in building resilience from Pakistan's perspective. Disasters increase vulnerability by damaging livelihood and physical assets, people, and the economy. The question arises what will be the benefit of implementing adaptive social protection to reduce vulnerability and increase resilience? The objectives of the research are to answer the following questions:

- To evaluate the existing Social Protection Programs (BISP, Ehsaas) in order to explore synergies between social protection, DRR, and CCA (ministry of climate change)
- To develop a framework for strengthening DRR through social protection and climate change adaptation (Case Study NDRMF).

The focus of the research was restricted to three organizations—Ehsaas Program/Benazir Income Support Program (BISP), the National Disaster Risk Management Fund (NDRMF), and the Ministry of Climate Change (MoCC). Their synergy is explored as per the concept of Adaptive Social Protection (ASP) given by Davies et al. (2013), and the relationship among thematic areas of three policies is established through Causal Loop Diagram (CLD).

2 DRR, SP and CCA and Their Integration with Pakistan Perspective

2.1 Disasters in Pakistan

Pakistan is having a population of 207.77 million (Statistics, 2017), and its geography comprises mountains (HKH), deserts (Thar, Cholistan, etc.), and plains (Indus river plains). More than 60% area receives less than 200 mm/annum rainfall and hence remains dry. Monsoons and heat waves cause flooding, and the collision of Eurasian and Indian tectonic plates causes seismic instability. Other hazards for Pakistan include landslides, river erosion, pest attacks, epidemics, tsunami, Glacial Lake Outburst Floods (GLOF), avalanches, cyclones, storms, technological accidents, and droughts. Human-induced hazards include oil spills, industrial, transport, urban, and forest fires. However, droughts, earthquakes, floods, landslides, and wind storms are high priority hazards (Khan, 2010). Pakistan is also ranked 8th out of the 10 most vulnerable countries to climate change effects (UNDRR, 2019), a disaster-prone country due to its geophysical location, and climatic and economic conditions and has suffered a loss of US $18 billion during the last decade due to disasters (World Bank, 2017). Per capita income is $1487 in the year 2018 which falls in the lower middle-income category. Disasters further aggravate the situation by having a grave impact on economic growth and human lives (Table 1).

Severe droughts of 1998–2001 reduced Gross Domestic Product (GDP) growth by 50%, and earthquake in 2005 and flooding in 2010, 2011, and 2012 resulted in an average economic loss of 1.16% of GDP (Ehsan, 2020). Asim Fayaz (July 2017) assessed loss and damage from three major disasters in Pakistan. The details are as follows:

Table 1 Poverty data of Pakistan

Poverty	Number of poor (in million)	Rate (%)	Period
National Poverty Line	46.0	24.3	2015
International Poverty Line 63.9 in Pakistani rupee (2015) or US$1.90 (2011 PPP) per day per capita	7.5	3.9	2015
Lower Middle Income Class Poverty Line 107.6 in Pakistani rupee (2015) or US$3.20 (2011 PPP) per day per capita	65.7	34.7	2015
Upper Middle Income Class Poverty Line 184.9 in Pakistani rupee (2015) or US$5.50 (2011 PPP) per day per capita	142.8	75.4	2015

Source World Bank, 2017

- The floods of 2010 resulted in a loss of $10 billion, and 20 million people were affected.
- Loss of $ 5.2 billion from the earthquake 2005.
- The drought of 1998–2000 affected 2.2 million population.

2.2 DRR

Foreign aid helped disaster response, recovery, and rehabilitation during all disasters, but the government of Pakistan has shared primary responsibility to invest in DRR for resilience as per the Sendai framework of action. Pakistan formulated a DRR policy to increase resilience through the reduction of vulnerability and future development/programs. Three focus areas of DRR policy are risk knowledge, prevention and mitigation, and preparedness (DRR Policy Pakistan, 2013). A dedicated fund National Disaster Risk Management Fund (NDRMF) has been established under the companies' ordinance to invest in DRR activities regarding disaster preparedness, mitigation, and early warning. NDRMF is to carry out its projects in line with DRR policy 2013, Climate Change Policy 2012, SDGs (Government Vision 2025), and Sendai framework of action.

2.3 SP

Disasters' data shows that the poor are hit hardest due to limited ownership of assets. Existing major institutions for the implementation of social assistance programs of SP in Pakistan are Benazir Income Support Program (BISP), Zakat, and Bait-ul-Maal.

Most recently "**Ministry of Social Protection and Poverty Alleviation Coordination (MSP & PAC)**" has been created to bring fragmented organizations including Trust for Voluntary Organizations, Pakistan Bait-ul-Mal, Benazir Income Support Program, Centre for Social Entrepreneurship *Zakaat*, Pakistan Poverty Alleviation Fund (PPAF), the SUN Network, and secretariats of the Poverty Alleviation Coordination Council and planned Labor Expert group under one umbrella for one window operation to reduce abuse of funds and duplication (Ehsaas Program).

Pakistan is also being affected due to global climate change because climate change has no geographical/political boundaries. An increase in precipitation and dry periods has been observed during the past two decades due to a gradual temperature rise. Climate change has disturbed the hydrological cycle which has resulted in a change in water availability periods, droughts, cropping patterns, precipitation patterns, frequency, and intensity of heat waves and floods (Ali et al., 2009). Global warming has a deep impact on the Hindu Kush-Karakoram-Himalayan (HKH) glaciers which are the major source of feeding Pakistan's rivers and are receding constantly due to global warming (IPCC, 2007). Pakistan's economy is agrarian, and its energy sector is largely dependent on water resources. Hence, its water, food, and energy resources are at risk. Climate Change Policy 2012 has been formulated for the conservation of water resources (enhancing capacity, awareness, and infrastructure development), agriculture, human health, forestry, biodiversity, and protection of vulnerable ecosystems (Climate Change Policy Pakistan, 2012). The Ministry of Climate Change (MoCC) is carrying out different Projects for Climate Change Adaptation in accordance with Climate Change Policy.

The building block of DRR policy 2013 is to increase resilience through vulnerability reduction. The purpose of SP programs (Ehsaas, BISP) is to help orphans, widows, poor, homeless, jobless, poor farmers, laborers, low-income students, poor women, and elderly citizens. The objective of Climate Change Policy 2012 is socioeconomic growth by addressing climate-related issues and pro-poor gender-sensitive adaptation. The concept of Adaptive Social Protection was developed to combine interventions and policies related to DRR, SP, and CCA (Davies et al., 2013). The core point for the conjunction of DRR, SP, and CAA is to bring resilience in society against disaster and shocks due to hazards and climate change. This concept of harmonized implementation of Social Protection Programs and DRR and Climate Change Policies will help to achieve community resilience and to mitigate/prevent the losses to elements at risk (people, assets, economic, and social) due to disasters and climate change.

3 Study Design

This research is multi-dimensional, and most of the data regarding these three communities working for the same cause was collected through a literature review and consultation with professionals in relevant fields. Literature review and data collection proved poverty is the root cause of vulnerability to disasters. So poverty

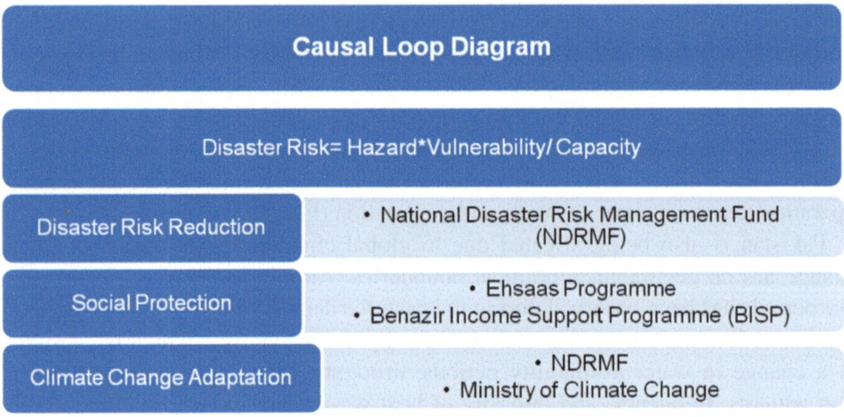

Fig. 1 The pattern of data collection

alleviation and resilience were taken as the baseline of the study. The scheme of research descends from a general literature review and surveys to prove Causal Loop Diagram (CLD) to the specific function of three fields in Pakistan. The role of Social Protection (Ehsaas Program), Disaster Risk Reduction (NDRMF), and Climate Change Adaptation (Ministry of Climate Change-MoCC) were researched to resolve poverty vulnerability and to build resilience. Policies of each field have been made part of the research for better understanding and implementation. Projects are shown in tabular form, and synergy is explored among three fields.

The pattern of the study is shown in Fig. 1.

3.1 Analysis Tools

Causal Loop Diagram (CLD) is used to map the thematic areas of three streams i.e., Social Protection, Disaster Risk Reduction, and Climate Change Adaptation Policies for poverty alleviation and building resilience in society. National Disaster Management Fund for implementation of Disaster Risk Reduction (DRR) Policy is taken as balancing (negative) reinforcing feedback loops' relationships, ministry of climate change and Social Protection for implementation of Climate Change Policy and Ehsaas Program are taken as reinforcing (positive) feedback loops relationships based on discussion with experts and questionnaires from the professionals. (McGlashan et al., 2016) described that Causal Loop Diagrams (CLD) help in mapping complex problems. CLD comprises variables, polarity, and causal relationships. Variable is the dynamics of cause and effect of the issue under consideration. The causal relationship is denoted by arrows in between variables. Polarity is the orientation of the relationship i.e., variables changing in the same direction have positive polarity while variables in opposite direction have negative polarity. Start with a simple CLD and create links and feedback between variables. We used literature and made use of your understanding to test CLD (Haraldsson, 2004). Richardson and Pugh III (1981) defined that a causal loop diagram consists of feedback loops. These loops are closed in nature to show cause and effect. Information and action are depicted through these loops. There are two types of causal loops—positive feedback (Reinforcing) and negative feedback loops (balancing) (Kirkwood, 1998). Positive feedback augments change further in the same direction of increase or decrease. Negative feedback has the opposite effect. This is a goal-seeking loop. The variable pushes back to the goal by increasing or decreasing in the direction of the goal if the variable lies below or above the goal.

4 Findings and Discussions

In this research, we used a combination of primary (interview and workshops) and secondary (statistics and research) data to develop a qualitative set of feedback loops known as a causal loop diagram (CLD) (Fig. 2).

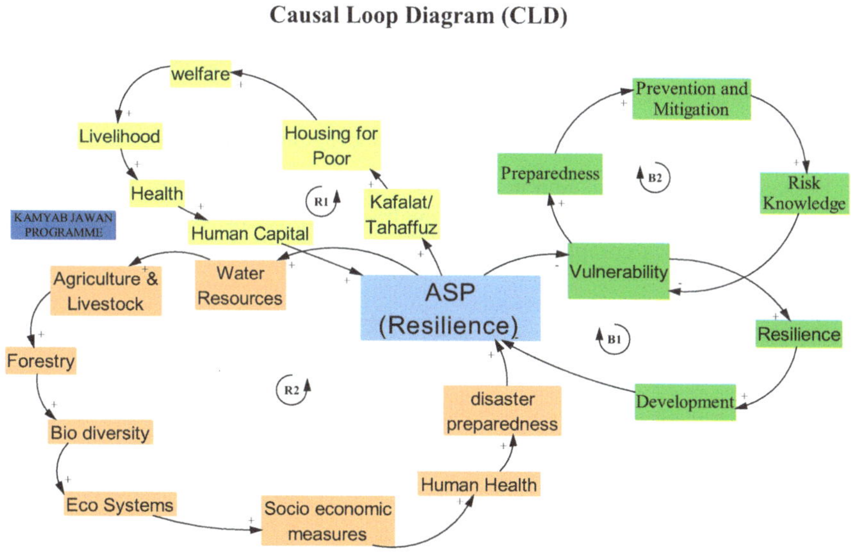

Fig. 2 Causal loop diagram generated based on Ehsaas Program, DRR, and Climate Change Policy

5 Role of Social Protection for Vulnerability Reduction

5.1 Re-enforcement (R1) Proof

Social protection and safety nets are recognized by government of Pakistan as a means to manage and mitigate risk. These are the tools to reduce poverty and hence vulnerability through insurance and transfers as mitigation to disasters (Survey, 2013–2014). Social safety nets are an important tool for poverty eradication and the improvement of health and education (Mumtaz & Whiteford, 2017).

Evidence shows that safety nets help nations to invest in human capital (World Bank). Social safety net and poverty alleviation policies should be inclined to invest in the health and education sector to eliminate human capital poverty (Hanmer, 1998). Some Conditional Cash Transfers (CCTs) are programs for poor households with the precondition of investment in the human capital development of children. These CCTs are provided for the nutrition and health of children (Fiszbein & Schady, 2009).

Robert Chambers and Gordon Conway (1992) defined livelihood as means of living through assets, capabilities, and activities. The five elements of livelihood are physical, natural, human, social, and financial (Dubey et al., 2015). People have central importance in livelihood approaches because of their ability to convert assets into livelihood. Human capital includes knowledge, skills, good health, and the ability to perform arduous jobs for achieving livelihood objectives (Fig. 3).

Thus, livelihood depends on the quality of human capital.

- **Ehsaas Program/Benazir Income Support Program**

Ehsaas Program was inaugurated by Prime Minister on 27 March 2019 with the main objective to reduce poverty and inequality. The principles of the program are the involvement of a complete government system, strengthening of institutions,

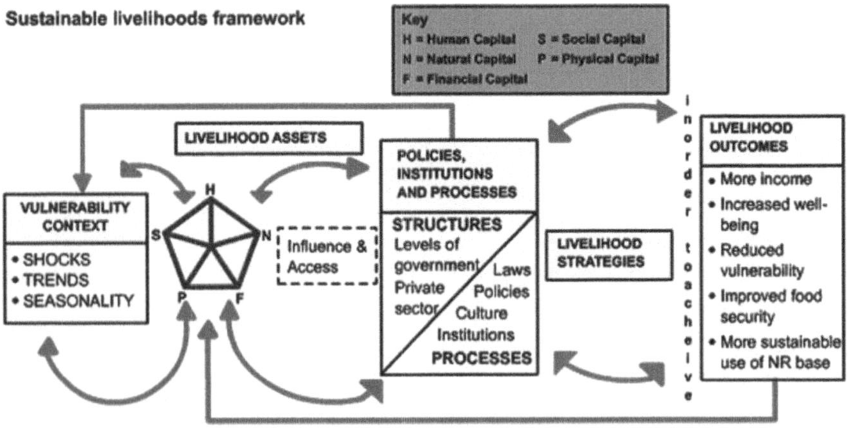

Fig. 3 Sustainable livelihood framework. *Source* DFID (1999)

Table 2 Projects carried out under Ehsaas Program

Thematic area	Project	Details
Safety Net	Kafalat/Tahaffuz	Rs. 12,000/—quarterly cash transfer to poor households. COVID emergency relief program to support the unemployed in addition to existing beneficiaries Benazir Income Support Program
	Naya Pakistan Housing Scheme	
Human Capital	Nutrition	School projects, kitchen gardening
	Education	Scholarships CCT
	Health	Insaaf Cards
Livelihood		Youth skill development program Microfinancing to poor and laborers for tree plantation, olive oil plantation, and nurseries' growth under the billion tree tsunami program

good governance, and involvement of the private sector. The beneficiaries of the program are orphans, extremely poor, widows, disabled, homeless, undernourished, medical problems, laborers, needy students, poor farmers, elderly citizens, and poor women. Ministry of Social Protection and Poverty Alleviation has been created to bring fragmented social protection institutions (Zakat, Bait-ul-Maal, Benazir Income Support Program and Pakistan Poverty Alleviation) under one umbrella. The database has central importance for safety net programs like the Ehsaas Program, and the Benazir Income Support Program (BISP) for the identification of actual eligible beneficiaries and transparency of projects. BISP with the help of the World Bank conducted a national level survey in the year 2010–2011. Information on the social, economic, and welfare status of approximately 27 million households was collected through this survey (BISP). The National Socio-Economic Registry (NSER) has been linked with National Data Base and Registration Authority (NADRA). Projects being carried out by Ehsaas Program are as follows: (Table 2).

6 Role of Disaster Risk Reduction (DRR) for Vulnerability Reduction

6.1 Balancing Loop Proof

Four phases of disaster management elaborated by Poser and Dransch (2010) include disaster mitigation, preparedness, response, and recovery (Fig. 4).

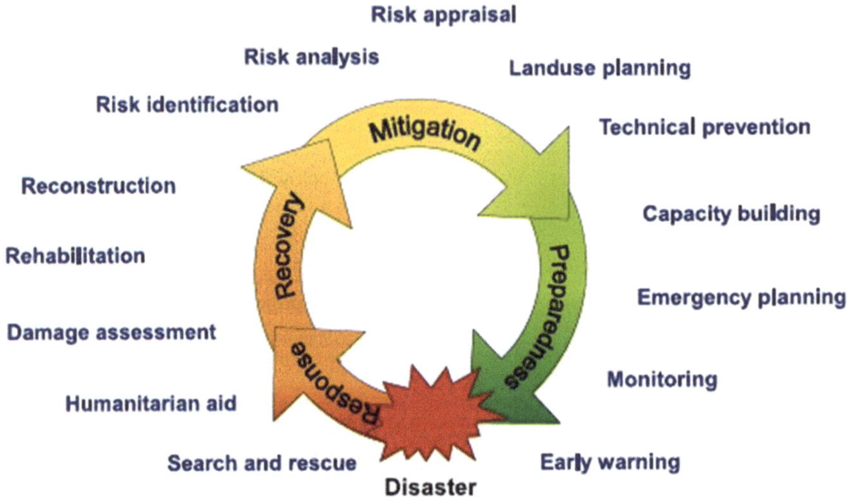

Fig. 4 The phases of the disaster management cycle and examples of related activities. *Source* Poser and Dransch (2010)

DRR policy of Pakistan (balancing loop) is incorporated in the sequence of this figure in the causal loop diagram.

- **National Disaster Risk Management Fund (NDRMF)**

Pakistan is a disaster-prone country due to its geophysical location. Pakistan is ranked 6 out of 10 hazard-prone countries as per the Index for Risk Management (INFORM) report 2018. Pakistan is number 8 out of the 10 most affected countries due to climate change (UNDRR, 2019). The earthquake of 2005 and the floods of 2010 resulted in colossal human and economic losses. Government realized the need to invest in DRR in order to mitigate the effects of disasters. Hence, National Disaster Risk Management Fund (NDRMF) was established under the companies' ordinance to invest in projects to enhance resilience against natural and climatic hazards and to reduce socioeconomic vulnerability. The objective of the fund is preparedness, mitigation, and early warning systems. The fund is not to carry out any post-disaster activity. DRR Policy 2013, Climate Change Policy 2012, National Flood Protection Plan, and Sendai Framework are the policies and guidelines to be followed by NDRMF. NDRMF has established a synergy group, comprising BISP, the Ministry of Finance, Commerce, and the Economic Affairs divisiòn, and members from the insurance industry and donors' representatives, for Disaster Risk Financing (DRF) interventions and coordination among stakeholders. The objective of the synergy group is to exchange information and to bring transparency to DRF projects. Preparation of Natural Catastrophe Model (NATCAT)—hazard, vulnerability, and expected losses mapping are in progress. Projects being carried out by NDRMF are as follows: (Table 3).

Table 3 Projects carried out under NDRMF

Thematic area	Projects
MHVRA	MHVRA in 20 vulnerable cities Vulnerability and risk assessments at local levels
CBDRM, mitigation, prevention and preparedness, climate change, early warning	Promoting integrated mountain safety in Northern Pakistan (PIMSNP) Mitigation and response through CBDRM in the 50 most vulnerable union councils
	From vulnerability to resilience (V2R) Vulnerability and risk assessments in 15 districts of Baluchistan and KPK
	Resilient and Adaptive Population in Disaster (RAPID)
	Physical infrastructure, CBDRM, and Community Emergency Response Teams
	Building resilience by strengthening the community through inclusive disaster risk management
	Building resilience to disasters and climate change
	Retrofitting 300 km of new or retrofitted flood protection infrastructure to be built
	Resistance to multiple hazards 500 social public sector buildings

7 Role of Climate Change Adaptation (CCA) for Vulnerability Reduction

Hydrometeorological events are being increased due to climate change. Rapid urbanization, population growth, and increase in asset values are the main drivers of climate change which increase the intensity and frequencies of water-related extreme events. Comprehensive actions are required to increase resilience and reduce the vulnerability of exposed assets and populations (Wieriks & Vlaanderen, 2015). Agriculture production and water management are interrelated. A major portion of Pakistan's economy is based on agriculture. Irrigation has a central role in agriculture as most of the agricultural production is dependent on irrigation. Agriculture provides livelihood to more than 66% population of Pakistan, 21% contribution to GDP, and 11% to exports. Although the economy is being shifted to urbanization and industrialization, the main focus will remain on the agriculture sector. An increase in the supply of water for irrigation is required for the increase in agricultural production (Haq & Shafique, 2009). Water is crucial for livestock and agriculture. Diversification of livestock and agriculture is also required to increase their adaptive capacity against future water scarcity (Naqvi et al., 2015).

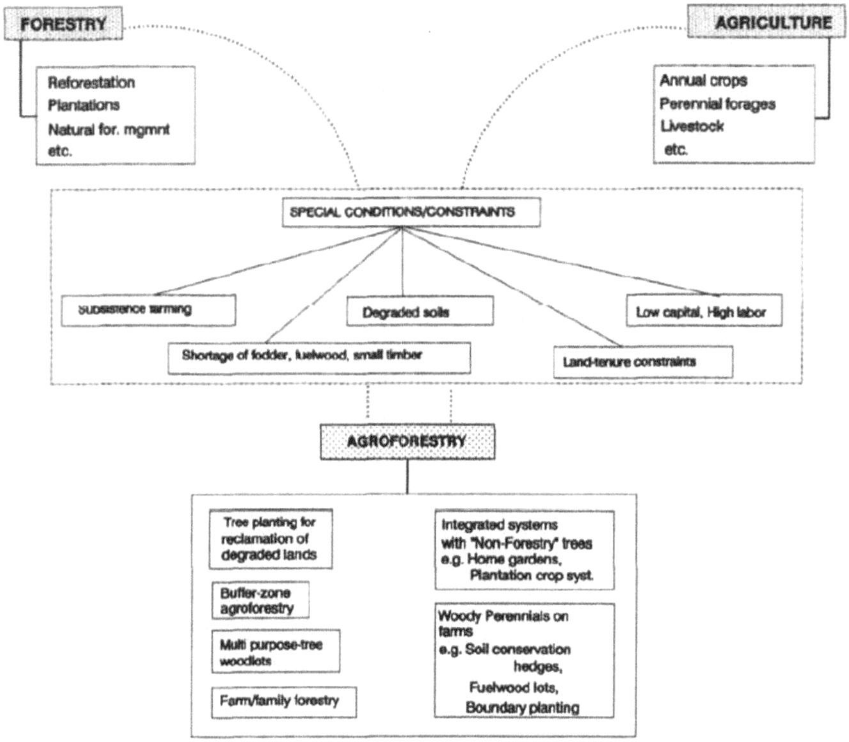

Fig. 5 Agroforestry has developed as an interface between agriculture and forestry. *Source* An introduction to agroforestry P. R. Nair (1993)

Agroforestry is defined as woody perennials and crops on the same land tract. An interaction between crops and forests fulfills economic and ecological needs, especially in developing countries (P. R. Nair, 1993). This relationship is mentioned in the following Fig. 5.

Land degradation is curtailed through the development of synergies between livestock and crops (Lemaire et al., 2014). Integration of forestry with livestock and crops brings in diversity (Alves et al., 2017). Reintroduction of local species, reforestation and conservation of species, and forests are needed to conserve biodiversity (Sodhi et al., 2010). Overlapping between biological conservation and poverty can be used as a win-win situation i.e., poverty eradication with the help of biological diversity and conservation of biodiversity by increasing the social status of the community (Fisher & Christopher, 2007) (Fig. 6).

Climate change has negative effects on health, and this impact is disproportionate to poor and developing countries due to their limited adaptation capacity. Assessment of vulnerability and adaptation strategies are needed to offset the impact of climate change on the health sector. Planning and mitigation strategies can reduce health issues. For example, air pollution can be reduced by utilizing renewable alternate

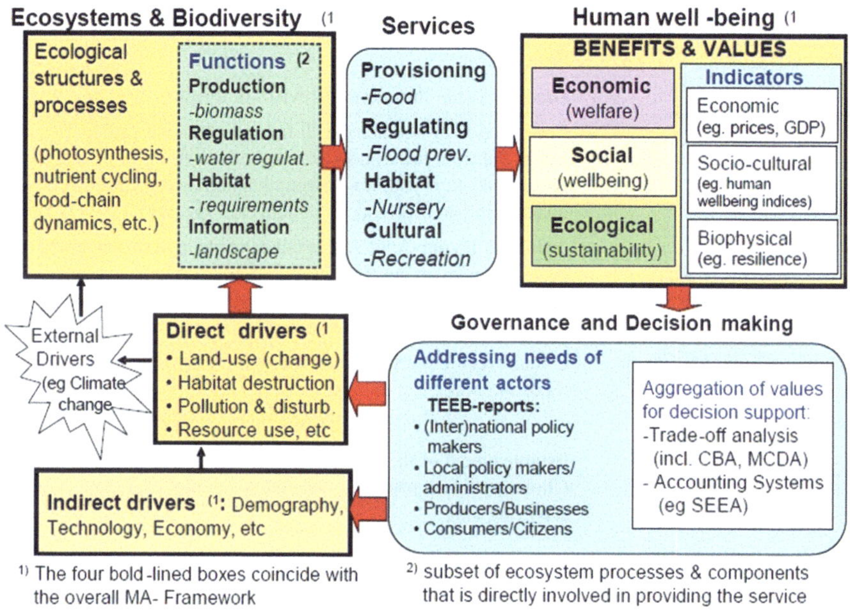

Fig. 6 A conceptual framework for linking ecosystems and human well-being. *Source* De Groot et al. (2010)

energy sources with ultimate improvement in health aspects (Haines et al., 2006). Income is the basic socioeconomic variable having a direct impact on health status. The health of low income, poor, elderly, and the unemployed are disproportionately affected. The mortality rate is associated with social class (Public Health Association, 1992). Vulnerable populations with chronic diseases, poor health, and disabilities have a high risk of serious health issues after natural disasters (Bethel et al., 2011). Actions were taken to build the resilience of a community before the occurrence of a disaster and augment disaster preparedness strategies (Cavallo, 2014). Projects being carried out by the Ministry of Climate Change are as follows: (Table 4).

8 Conclusions and Recommendations

8.1 Conclusion

This study has concluded that poverty is the root cause of vulnerability to disasters. Poverty puts pressure on people to live in unsafe conditions which makes them vulnerable to disasters. Poor are disproportionately affected by disasters. Social Protection is a tool to cope with poverty. DRR mainly deals with the reduction of vulnerability and increase in coping capacity. CCA deals with an increase in adaptive

Table 4 Projects carried out under MoCC

Thematic area	Project
Forestry	Green Pakistan Programs—Revival of forestry
Biodiversity	Construction of boundary wall for zoo/botanical garden Green Pakistan Programs—Revival of wildlife Green Pakistan Program—Strengthening zoological survey undertaking an inventory of endangered species
Eco systems	**Arid and Hyper-Arid Areas** (sustainable land management project to combat desertification in Pakistan) **Human Health** establishment of Pakistan WASH Strategic Planning and Coordination Cell
Socioeconomic Measures	Climate resilient urban human settlement units; Climate change reporting unit
Human health	
Disaster preparedness	Establishment of the geomatics center for climate change Climate resilient urban human settlement units; Climate change reporting unit

capacity against the adverse impacts of climate change and a reduction in vulnerability against climate-related extreme events like floods, droughts, cyclones, heat waves, wind storms, etc. In Pakistan, Ehsaas Program provides social protection for poverty alleviation, NDRMF carries out projects in the field of DRR and CCA to reduce vulnerability to disasters, and MoCC looks after CCA in accordance with Climate Change Policy. A synergy group exists in NDRMF (members from diversified fields including reps of BISP and MoCC) for better coordination between social protection, CCA, and DRR. Regular meetings are held by the synergy group. Ehsaas Program and NDRMF contributed to the COVID-19 relief fund.

8.2 Recommendation

- Database plays a central role in prevention, mitigation, and response actions. NSER data is available and the NATCAT model is used to prepare the hazard, vulnerability, and losses map which is in progress. Synergy group is established by NDRMF for coordination among stakeholders. The same forum may be used for data and experience sharing/coordination among stakeholders.
- Climate Change Adaptation and Disaster Risk Reduction projects in Pakistan may ensure Social Protection through local employment through public works, micro credit/microfinance, insurance schemes for the poor, gender representation, capacity building of local institutions, protecting assets, and livelihood diversification.

- Insurance coverage for the poor segment of society is a weak area because of their inability to pay the premium. NDRMF may pay a premium for insurance coverage for poor.

References

Akter, S., & Mallick, B. (2013). The poverty–vulnerability–resilience nexus: Evidence from Bangladesh. *Ecological Economics, 96*, 114–124.

Ali, G., Hasson, S., & Khan, A. M. (2009). Climate change: Implications and adaptation of water resources in Pakistan. *Global Change Impact Studies Centre (GCISC): Islamabad, Pakistan.*

Alves, B., Madari, B., & Boddey, R. (2017). Integrated crop–livestock–forestry systems: Prospects for a sustainable agricultural intensification. *Nutrient Cycling in Agroecosystems, 108*, 1–4. https://doi.org/10.1007/s10705-017-9851-0

Asim Fayaz, J. B. (2017, July). *Disaster preparedness in Pakistan.* www.strausscenter.org›cepsa-research-briefs›download=649:disast.

Association, P. H. (1992). *The impact of economic and social factors on health* (Report prepared by the Public Health Association of New Zealand, Wellington).

Bethel, J. W., Foreman, A. N., & Burke, S. C. (2011). Disaster preparedness among medically vulnerable populations. *American Journal of Preventive Medicine, 40*(2), 139–143.

BISP. *National Socio Economic Registry.* https://bisp.gov.pk/nser/

Cavallo, A. (2014). Integrating disaster preparedness and resilience: A complex approach using system of systems. *The Australian Journal of Emergency Management, 29*(3), 46.

Chambers, R., & Conway, G. (1992). *Sustainable rural livelihoods: Practical concepts for the 21st century.* Institute of Development Studies (UK).

Davies, M., Béné, C., Arnall, A., Tanner, T., Newsham, A., & Coirolo, C. (2013). Promoting resilient livelihoods through adaptive social protection: Lessons from 124 programmes in South Asia. *Development Policy Review, 31*(1), 27–58.

De Groot, R. S., Fisher, B., Christie, M., Aronson, J., Braat, L., Haines-Young, R., Gowdy, J., Maltby, E., Neuville, A., Polasky, S., & Portela, R. (2010). Integrating the ecological and economic dimensions in biodiversity and ecosystem service valuation. In *The Economics of Ecosystems and Biodiversity (TEEB): Ecological and economic foundations* (pp. 9–40). Earthscan, Routledge.

DFID. (1999). *Sustainable livelihood framework.* Department for International Development.

Dubey, P., Bhadra, B., & Santra, S. (2015). Role of human capital for changing livelihood pattern: A case study in Nadia District of West Bengal. *International Research Journal of Social Sciences, 4*(4), 98–104.

Ehsan, U. (2020). *Exploring synergies between social protection programs, disaster risk reduction (DRR) and climate change policies in Pakistan: Case Study NDRMF*, MS Thesis, National University of Sciences and Technology, Pakistan.

Engle, N. L., de Bremond, A., Malone, E. L., & Moss, R. H. (2014). Towards a resilience indicator framework for making climate-change adaptation decisions. *Mitigation and Adaptation Strategies for Global Change, 19*(8), 1295–1312.

Fisher, B., & Christopher, T. (2007). Poverty and biodiversity: Measuring the overlap of human poverty and the biodiversity hotspots. *Ecological Economics, 62*(1), 93–101.

Fiszbein, A., & Schady, N. R. (2009). *Conditional cash transfers: Reducing present and future poverty.* The World Bank.

Haines, A., Kovats, R. S., Campbell-Lendrum, D., & Corvalán, C. (2006). Climate change and human health: Impacts, vulnerability and public health. *Public Health, 120*(7), 585–596.

Haraldsson, H. (2004). *Introduction to system thinking and causal loop diagrams.*

Hanmer, L. C. (1998). Human capital, targeting and social safety nets: An analysis of household data from Zimbabwe. *Oxford Development Studies, 26*(2), 245–265. https://doi.org/10.1080/13600819808424155

Haq, R., & Shafique, S. (2009). Impact of water management on agricultural production.

Heltberg, R., Siegel, P. B., & Jorgensen, S. L. (2009). Addressing human vulnerability to climate change: Toward a "no-regrets" approach. *Global Environmental Change, 19*(1), 89–99.

IPCC. (2007). *Assessment report.*

Khan, I., Ali, A., Waqas, T., Ullah, S., Ullah, S., Shah, A. A., & Imran, S. (2022). Investing in disaster relief and recovery: A reactive approach of disaster management in Pakistan. *International Journal of Disaster Risk Reduction, 75*, 102975.

Kirkwood, C. W. (1998). System dynamics methods. *College of Business Arizona State University USA.*

Lemaire, G., Franzluebbers, A., De Faccio Carvalho, P., & Dedieu, B. (2014). Integrated crop–livestock systems: Strategies to achieve synergy between agricultural production and environmental quality. *Agriculture, Ecosystems & Environment, 190*, 4–8. https://doi.org/10.1016/j.agee.2013.08.009

McGlashan, J., Johnstone, M., Creighton, D., de la Haye, K., & Allender, S. (2016). Quantifying a systems map: Network analysis of a childhood obesity causal loop diagram. *PloS One, 11*(10).

Mumtaz, Z., & Whiteford, P. (2017). Social safety nets in the development of a welfare system in Pakistan: An analysis of the Benazir Income Support Programme. *Asia Pacific Journal of Public Administration, 39*, 16–38. https://doi.org/10.1080/23276665.2017.1290902

Nair, P. R. (1993). *An introduction to agroforestry*. Springer Science & Business Media.

Naqvi, S., Kumar, D., De, K., & Sejian, V. (2015). Climate change and water availability for livestock: Impact on both quality and quantity. In *Climate change impact on livestock: Adaptation and mitigation* (pp. 81–95). Springer.

Government of Pakistan (2012). Climate Change Policy.

Government of Pakistan. (2013). DRR policy.

Poser, K., & Dransch, D. (2010). Volunteered geographic information for disaster management with application to rapid flood damage estimation. *Geomatica, 64*, 89–98.

Richardson, G. P., & Pugh III, A. I. (1981). *Introduction to system dynamics modeling with DYNAMO.* Productivity Press Inc.

Smit, B., & Pilifosova, O. (2003). Adaptation to climate change in the context of sustainable development and equity. *Sustainable Development, 8*(9), 9.

Sodhi, N. S., Koh, L.P., Clements, R., Wanger, T. C., Hill, J. K., Hamer, K. C., Clough, Y., Tscharntke, T., Posa, M. R. C., & Lee, T. M. (2010). Conserving Southeast Asian forest biodiversity in human-modified landscapes. *Biological Conservation, 143*(10), 2375–2384.

Statistics, P. B. (2017). *Annual report.* Pakistan Bureau of Statistics.

Survey, P. E. (2013–2014). *Pakistan economic survey.*

Tandon, A., & Hasan, R. (2005). Conceptualizing and measuring poverty as vulnerability: Does it make a difference?

UNDRR. (2019). Status report.

Vakis, R. N. (2006). *Complementing natural disasters management: The role of social protection*: World Bank, Social Protection.

Wieriks, K., & Vlaanderen, N. (2015). Water-related disaster risk reduction: Time for preventive action! Position paper of the High Level Experts and Leaders Panel (HELP) on water and disasters. *Water Policy, 17*(S1), 212–219.

World Bank. (2017). *Safety nets.* https://www.worldbank.org/en/topic/safetynets

Indigenous Knowledge and Flood Resilience Strategies in African Coastal Cities: From Practice to Policy

Olasunkanmi Habeeb Okunola, Kehinde Balogun, and Deborah Bunmi Ojo

1 Introduction

Flooding continues to be a severe type of climate risk affecting every African region, with varying social, economic, and environmental effects across continents. The World Bank (2017) and Rentschler and Salhab (2020) estimate that sub-Saharan Africa has at least 71 million people at risk of flooding. For instance, the International Federation of Red Cross and Red Crescent Societies [IFRC] (2021) is of the view that in 2020, small and moderate floods displaced nearly 7 million people and caused 1,273 deaths in 15 African nations. Climate change is predicted to worsen the situation in African coastal cities by making flood events more often, intense, and unprecedented (Intergovernmental Panel on Climate Change [IPCC], 2021). O'Donoghue et al. (2021) predict that African coastal cities will be the hardest hit by frequent flooding due to their low elevation and topography with less resilience. Several African coastal regions are currently susceptible to erosion and flooding, resulting in many ecological issues and a great deal of suffering. Realizing these consequences of flooding has prompted stakeholders in various nations to develop national strategies for managing frequent flood occurrences. Therefore, understanding flood resilience strategies in sub-Saharan African nations is of great significance and interest.

The concept of resilience has grown in popularity and now encompasses social, physical, economic, and institutional elements. Building flood resilience in this study represents enhancing the preparedness of nations, communities, and regions to deal with and swiftly implement recovery plans from frequent flood events (Patel et al.,

O. H. Okunola (✉) · K. Balogun
United Nations University—Institute for Environment and Human Security (UNU-EHS), Bonn, Germany
e-mail: olasunkanmi.habeeb@gmail.com

D. B. Ojo
Department of Urban and Regional Planning, Obafemi Awolowo University, Ile-Ife, Nigeria
e-mail: dbojo@oauife.edu.ng

N. Shahzad (ed.), *Water and Environment for Sustainability*,
https://doi.org/10.1007/978-3-031-27280-6_10

2020). According to Mannakkara and Wilkinson (2015), a resilient community can "build back better," which implies a community that is capable of recovering from various forms of extreme events with little or no outside support. Indigenous communities occupy an increasingly unique position in disaster resilience "thinking," as indicated in the Sendai Framework for Action 2015, as part of its call for governments to employ a people-centered approach and engage directly with indigenous people in building the disaster resilience of nations and communities. According to the Fifth Assessment Report (AR5) of the Intergovernmental Panel on Climate Change (IPCC), the integration of several pieces of knowledge into climate change research is key to the successful implementation of climate mitigation, and adaptation strategies. Additionally, indigenous knowledge and strong links to "land" necessitate unique attention. Where place or land broadly refers to environments to which cultural groups or individuals are linked, as well as situations that give worth and meaning to people's quality of life (Garnett et al., 2018), therefore, the cascading impacts of climate change on environmental and interpersonal interactions, life experience, spiritual connections, kinship, family, and oral history are frequently as significant as, or even more significant than, the more direct effects (Ford et al., 2020).

People in the global north and south countries have adapted their lives and livelihoods to changing environments over time by incorporating indigenous knowledge into their traditional disaster management systems (Alessa et al., 2016). In addition, studies such as Enock (2013) and Murdoch et al. (2020) argue that the utilization of indigenous knowledge based on the observations and experiences of community members represents the best way to assess climate change and its direct and indirect impacts. This knowledge may offer helpful insight into climate change prevention, mitigation, and preparedness. For instance, Kalanda-Joshua et al. (2011) established that farmers in southern Malawi make agricultural and farming decisions using traditional knowledge. According to the authors, the most prominent parameters used by farmers in the region are the lengthening of the rainy season (which indicates the onset of a cold season), the behavior of specific insects and birds, and the activity of specific patterns of wind that signal the end of the rainy season.

Recent research on indigenous knowledge and climate change has primarily centered on local observations of climate and environmental changes in rural areas of North America, the Arctic, the Polar, and the Tropics, Asia, and even Africa (Nakashima & Krupnik, 2018; Petzold et al., 2020; Savo et al., 2016). In coastal cities, however, little has been done to incorporate indigenous knowledge into research on flood risk reduction. In light of the growing evidence that effective adaptation to climate change depends on integrating scientific and local knowledge systems, this chapter addresses this knowledge gap by assessing flood risks and flood resilience strategies in African coastal cities. In addition, the chapter investigates the definition, the conceptualization of features and the purpose of indigenous knowledge in coastal cities, and the level of integration of indigenous knowledge and scientific knowledge, if at all, in African coastal towns.

2 Flood Risk in Sub-Saharan African Cities

Flooding continues to be one of the most prevalent natural disasters in the sub-Saharan African (SSA) region, placing a heavy burden on vulnerable communities, ecosystems, livelihoods, and critical infrastructures. In recent years, cities in sub-Saharan Africa (SSA) have experienced an increase in flood risk due to extreme rainfall events, increasing urbanization, the expansion of informal settlements, and the lack of planned water and sanitation infrastructure. According to EM-DAT (2017), there were 130 flood incidents in sub-Saharan Africa between 2007 and 2017, affecting nearly 15.6 million people. On the KwaZulu-Natal coast of South Africa in 2022, heavy rains, flooding, and subsequent mudslides reportedly killed approximately 500 people and damaged approximately 2,000 houses and 4,000 "informal" homes or shacks (Okunola, 2022).

Numerous studies have determined that many countries in sub-Saharan Africa are currently experiencing pluvial flooding from torrential rain, groundwater flooding, flooding from overflowing drains and channels with insufficient capacity to evacuate all incoming water, and flooding from major rivers (Douglas, 2017). In some informal settlements of the region, such as Umlazi in Durban, Agbogbloshie in Accra, and Mile 12 in Lagos, the combination of high tides and heavy rain is a common occurrence, disrupting economic activities, causing traffic congestion on the main roads, and restricting the movement of people from other parts of the cities (Douglas, 2017; Karley, 2009; Olajuyigbe et al., 2012). In 2020, pluvial and fluvial flooding in Sudan affected 2,006,000 people, destroyed 181,500 homes, and resulted in 417 deaths in the country's coastal communities (Elagib et al., 2021). Coastal cities in South Africa, Nigeria, Niger, Guinea, and Senegal, as well as many others in the region, have experienced a series of flooding incidents in 2022.

The risks associated with flood disasters in African cities have been significantly amplified by human activities (Abiodun et al., 2017). As noted by Eguaroje et al. (2015) and Salami et al. (2017), flood risk is primarily caused by human activities such as informal settlements on low-lying floodplains, unregulated urban growth, poor-quality drainage systems, and continuous urbanization. While around 52.13% of Africa's population currently resides predominantly in rural regions, the continent's urban population is anticipated to expand from 678 million in 2018 to approximately 1.5 billion by 2050, from 678 million in 2018 (UN, 2018). It is anticipated that the frequency, intensity, and trend of flood events will continue to rise in the region due to human-induced changes in river hydrology and land use, increased development in flood-prone areas, and climate change. In addition, the potential impacts of climate change include an increased risk of pluvial, fluvial, and coastal flooding in Africa (Feka & Ajonina, 2011; MacLeod et al., 2021), indicating the need for sustainable flood risk reduction.

Despite multiple calls to integrate flood risk adaptation plans with sustainable development (Fuldauer et al., 2022; Fuso Nerini et al., 2019; Sayers et al., 2013; Zhenmin & Espinosa, 2019), many African countries' Paris Agreement adaptation plans rarely mention their contribution to the sustainable development goals. The extensive environmental, social, and economic impacts of flood risk in Africa necessitate urgent and strategic approaches to flood risk reduction to achieve sustainable development objectives (Abiodun et al., 2017). It is essential to base policy and action decisions on the greatest available scientific and indigenous knowledge in the face of global climate change and its new difficulties, unknowns, and uncertainties (Mutambisi et al., 2020). Consequently, it is crucial to evaluate African nations' national adaptation and resilience plans for reducing flood risk and contributing to Sustainable Development Goals.

3 Strategies for Flood Resilience in Sub-Saharan African Cities

3.1 Concept of Resilience

In recent years, resilience has gained significance in promoting proactive responses to extreme events. Diverse academic disciplines, including physics, geography, environmental science, risk management, and engineering, have discussed resilience and its operationalization. According to Bhamra et al. (2011), resilience is a multidimensional and multidisciplinary concept used in various fields and disciplines as a protective mechanism to achieve developmental goals, recover from adversity, and maintain competence following trauma. The concept of resilience can be traced back to the fourth century (Manyena, 2014; Talubo et al., 2022). From there, the concept of resilience has undergone a lengthy etymological journey and has been adopted in various sources of stress, including poverty, conflict, diseases, corruption, environmental degradation, resource scarcity, and disasters. Regardless of the various meanings and definitions of resilience, it has become increasingly evident that resilience describes how individuals, communities, neighborhoods, and institutions positively adapt to significant and protracted sources of stress or sudden-onset shocks.

Resilience refers to the capacity to survive and adapt to hazards, prevent or reduce losses, and recover with minimal disruptions in hazard and disaster study domains (Adger et al., 2011; Cutter et al., 2008). The Department of International Development (DFID) defines resilience as the capacity of households, communities, and countries to manage change by transforming or maintaining living standards in the face of disasters without jeopardizing their long-term prospects (DfID, 2021). Using a similar line of reasoning, Okunola and Olawuni (2022) assert that resilience involves pre-disaster and post-disaster measures that minimize and prevent disaster-related losses, damage, and impacts. From these submissions, it can be deduced that resilience includes critical elements such as context, disturbance, capacity, and reaction, which determine local and national resilience.

According to Gardner and Dekens (2007), community resilience to disaster is determined by processes and societal structures within both global and local contexts of hazard and is essentially the outcome of practices, beliefs, and forms of knowledge.

In addition, the Sendai Framework for Disaster Risk Reduction 2015–2030 stresses the significance of integrating indigenous knowledge and practices with contemporary scientific knowledge in climate change adaptation and disaster risk assessments. Indigenous knowledge and practices in this context refer to human knowledge and abilities utilized for disaster risk reduction techniques. Similarly, several studies have demonstrated that successful risk reduction processes are founded on incorporating traditional and local community knowledge (Kurnio et al., 2021; Rahman et al., 2017). In addition, traditional knowledge serves as the foundation for local communities coping strategies, raises awareness, and strengthens community participation and cohesion (Iloka, 2016). Given the preceding, it is clear that indigenous knowledge is an essential component of disaster resilience in communities. The problem, however, is the departed focus of previous studies on assessing indigenous knowledge practices in flood resilience thinking in Africa. Therefore, this study holistically evaluates indigenous knowledge strategies for flood risk reduction in African coastal cities.

4 Indigenous Knowledge, Adaptation Techniques, and Flood Resilience in African Coastal Cities

Over the years, the severe impacts of extreme flood events on the critical infrastructure, livelihoods, sectors, and systems of African coastal cities have increased the importance of indigenous knowledge in the area (Ajayi & Mafongoya, 2017; Codjoe et al., 2014; Mapfumo et al., 2016). In African societies, indigenous methods are considered integral coping or resilience mechanisms for floods. This could be attributed to the fact that they are characterized by innovation, adaptation, and experimentation, are non-formal and transmitted orally, are culturally and contextually specific, and are closely tied to the survival and subsistence of local communities. Therefore, diverse groups and scholars have documented various indigenous strategies adopted by local communities in sub-Saharan African coastal cities (Hooli, 2016; Mavhura et al., 2013; Okunola, 2022). A summary of these strategies indicates that indigenous flood knowledge, which is socioeconomically and culturally oriented, increases people's resilience to flooding hazards and other extreme events.

Mavhura et al. (2013) examined indigenous coping strategies in two Zimbabwean villages prone to flooding in the Muzarabani district. The study identified various indigenous preventative and mitigating measures, such as avoiding construction materials prone to cracking during flooding and erecting protective barriers around a home. Other methods include storing household items and livestock on mud shelves constructed in the kitchen or outdoors and choosing crop varieties suited to the soil conditions, local climate, and timing of floods. However, the study concluded that indigenous strategies are only effective when floods are of low magnitude. Kasei et al. (2019) found that residents of Accra, Ghana's informal urban settlements, rely on various indigenous knowledge indicators to forecast climate hazards. For example, early warm temperatures indicate a favorable rainy season in the community. In addition, the appearance of the Avakpo bird, a long-legged frog with brownish camouflage skin, and green and brown grasshoppers indicate a rainy season. The study concluded that there is insufficient consideration and integration of indigenous knowledge in the government's primarily technical and scientific early warning programs. Using the same line of reasoning, Hooli (2016) identified indigenous knowledge as a key strategy for local communities in Northern Namibia for flood prediction, prevention, preparedness, and coping. As a result of rapid urbanization, population growth, a lack of government support, and the disappearance of many of the familiar trees, anthills, and relocation sites upon which indigenous knowledge had been based, only a tiny percentage of local communities applied indigenous-based methods to flood management, according to the study.

Similarly, Dintwa et al. (2022) found that the exchange of goods and services between community members increases the resilience of coastal communities in Ngamiland West District, Botswana, to extreme climate events. For example, it was reported that people exchanged their farm cultivating tools for goats: those who received goats used them for milk production. In contrast, those who received cultivating tools used them for farm production to improve their standard of living and assist in adapting to the effects of climate change. In addition, traditional knowledge of plants is utilized to preserve food and medicinal herbs. Obi et al. (2021) believed that indigenous knowledge served as flood forecasting, control, emergency, and coping strategies in the coastal communities of Delta State, Nigeria. The authors argued that indigenous strategies are more effective and play a significant role in mitigating the effects of flooding on coastal communities. Relatedly, Okunola and Olawuni (2022) observed that the big stone (locally known as LapanGwagwan) in the river Chachanga serves as an indigenous early warning strategy for frequent flooding and plays a crucial role in numerous aspects of the well-being of the indigenous community in Lapai Gwari, Nigeria. Elders in Lapai Gwari explained, for instance, that the LapanGwagwan stone is used to determine the frequency of flooding and to categorize rainfall intensity as light, moderate, heavy, or violent rain. A heavy or violent downpour in the community is a sign that severe flooding is imminent.

Apraku et al. (2018) identified various factors for protecting animals and water bodies in South Africa's Eastern Cape to conserve biodiversity. Due to their medicinal and cultural importance to the Xhosa community, animals and waterways are protected in the Eastern Cape of South Africa. For instance, frogs are protected because they herald the beginning of the farming season and invite rain. Furthermore, when they cannot afford clinical treatment, community members use plants as a traditional medicine to treat diseases resulting from flood risk. In Tanzania, Boerhavia, conyzoides, Latana camara, Combretum collinum, Ageratum, Caparis tomentosa, and Biden Pilosa have been prioritized for conservation due to their potential medicinal value, according to Selemani (2020). According to ancient beliefs, plants also represent the relationship between the earth and the sky (da Silva et al., 2019). It is evident from these discussions that indigenous knowledge serves as preventative measures, adaptive measures, and distress mitigation for frequent flooding in sub-Saharan African coastal communities.

5 Indigenous Resilience Theory to Practice

Policymakers, environmental scientists, and community planners create and implement theories to address extreme climate-related events escalating and systemic nature. Various alternative solutions are utilized to address environmental conditions, and new models are continually being developed. Nevertheless, most of these models and innovations continue to exclude indigenous knowledge practices from global decision-making and policymaking processes. Even though international programs such as Rockefeller's 100 Resilient Cities and Sendai Framework for Disaster Risk Reduction 2015–2030 emphasize the importance of indigenous, traditional, and local knowledge in mitigating disaster risk, this is not the case in the United States. Indigenous or traditional knowledge can improve cost-effective, participatory, and sustainable adaptation strategies, according to the most recent IPCC assessment (AR6, published in 2022).

Despite the numerous hypotheses regarding the significant roles indigenous knowledge plays in the adaptation to extreme events, much work remains in terms of implementation and practice. This could be attributed to problems with the influence of religion and education, lack of proper documentation, knowledge transfer, dissemination, lack of recognition of forecasters, and lack of indigenous and scientific knowledge integration to adapt to climate change. In addition, research by Mavhura et al. (2013) and Tengö et al. (2014) indicates that indigenous coping strategies tend to have a limited impact on minimizing damage in extreme cases.

In light of the foregoing, the new models and innovations for resilience practice should recognize indigenous knowledge and strive for the authentic integration of scientific knowledge and indigenous knowledge toward the resilience of compounding extreme events. The starting point is the utilization of community-based participatory methods in assessing patterns of resilience (Bardosh et al., 2017), followed by the inclusion of traditional methods and the discovery of new and strategic ways to build and utilize partnerships among on-the-ground practitioners, policymakers, and academics. In addition, it is essential to recognize the significance of regional context and the fact that no "one-size-fits-all" model can address every region's extreme event (Korovulavula et al., 2020). Following this, the new indigenous resilience practice must consider contextual differences such as cultural, linguistic, jurisdictional, and governmental capacity and risk perception differences when designing the "best" resilience program and implementing best practices for each location.

By incorporating it into modern technologies and management practices, indigenous climate knowledge can also be used to enhance scientific forecasts and measures and increase local buy-in (IPBES, 2020; Leal Filho et al., 2022). This transparency allows local communities to collaborate with scientists and develop forecasts and adaptation strategies consistent with their practices and expectations. To establish novel collaborative arrangements, policymakers, climate scientists, and local communities must rely on participatory learning and outreach. In this process, not only will scientists make recommendations to policymakers for the improvement of community resilience strategies, but local communities will also have the opportunity to participate in planning and implementation stages and increase their capacity to facilitate a sustainable and inclusive indigenous resilience practice (Leal Filho et al., 2022; Nyong et al., 2007).

6 Discussion

This chapter identified various types of indigenous knowledge and its potential for enhancing community resilience to the accumulative and systemic risks posed by flooding and other extreme events in African coastal cities. In line with this, numerous African nations have adopted indigenous flood forecasting, food preservation, traditional medicine, and disaster preparedness and response strategies. Its influence in Ghana, Nigeria, Uganda, South Africa, and Zimbabwe is expanding. However, the dynamic transmission of indigenous knowledge in these nations is threatened by inadequate knowledge transfer and a lack of recognition from scientists and policymakers. Consequently, the applicability of indigenous methods to climate risks is highly localized and disconnected from scientific knowledge in the region, and many of these methods have recently vanished due to rapid societal change. It was determined, for instance, that indigenous knowledge practice in northern Namibia is being eroded due to rapid population growth, urbanization, a lack of government support, and the methods perceived ineffective in the face of severe flooding. Similarly, in many African nations, the conventional or scientific approach to extreme events has not met expectations. This chapter argues that one way to improve and strengthen Africa's resilience to extreme events is to complement scientific and indigenous knowledge with novel collaborative arrangements to boost capacity development in African coastal cities.

Studies such as McNamara and Prasad (2014), Madhanagopal and Pattanaik (2020), and Iwama et al. (2021) indicate that indigenous knowledge is also widely practiced in other regions, such as Asia, Latin America, and the Pacific region, where it is increasingly combined with scientific knowledge to gain a better understanding of climate change adaptation slow-onset effects. Combining indigenous and scientific knowledge could improve service delivery, cost-effectiveness, and success while encouraging indigenous communities to assume responsibility for future interventions and actively participate in them. It also encourages innovative scientific thought. Similarly, indigenous knowledge is fundamental to culture, heritage, indigenous identity, languages, and ways of life, and its transmission from one generation to the next must be supported, safeguarded, and preserved. Consequently, local and indigenous knowledge should be integrated into policies and mechanisms to mitigate the effects of climate change in African coastal cities.

References

Abiodun, B. J., Adegoke, J., Abatan, A. A., Ibe, C. A., Egbebiyi, T. S., Engelbrecht, F., & Pinto, I. (2017). Potential impacts of climate change on extreme precipitation over four African coastal cities. *Climatic Change, 143*(3), 399–413.

Adger, W. N., Brown, K., Nelson, D. R., Berkes, F., Eakin, H., Folke, C., Galvin, K., Gunderson, L., Goulden, M., O'Brien, K., Ruitenbeek, J., & Tompkins, E. L. (2011). Resilience implications of policy responses to climate change. *Wiley Interdisciplinary Reviews: Climate Change, 2*(5), 757–766.

Ajayi, O. C., & Mafongoya, P. L. (2017). *Indigenous knowledge systems and climate change management in Africa*. CTA.

Alessa, L., Kliskey, A., Gamble, J., Fidel, M., Beaujean, G., & Gosz, J. (2016). The role of indigenous science and local knowledge in integrated observing systems: Moving toward adaptive capacity indices and early warning systems. *Sustainability Science, 11*(1), 91–102.

Apraku, A., Akpan, W., & Moyo, P. (2018). Indigenous knowledge, global ignorance? Insights from an Eastern Cape climate change study. *South African Review of Sociology, 49*(2), 1–21.

Bardosh, K. L., Ryan, S. J., Ebi, K., Welburn, S., & Singer, B. (2017). Addressing vulnerability, building resilience: Community-based adaptation to vector-borne diseases in the context of global change. *Infectious Diseases of Poverty, 6*(1), 1–21.

Bhamra, R., Dani, S., & Burnard, K. (2011). Resilience: The concept, a literature review, and future directions. *International Journal of Production Research, 49*(18), 5375–5393.

Codjoe, S. N. A., Owusu, G., & Burkett, V. (2014). Perception, experience, and indigenous knowledge of climate change and variability: The case of Accra, a sub-Saharan African city. *Regional Environmental Change, 14*(1), 369–383.

Cutter, S. L., Barnes, L., Berry, M., Burton, C., Evans, E., Tate, E., & Webb, J. (2008). A place-based model for understanding community resilience to natural disasters. *Global Environmental Change, 18*(4), 598–606.

da Silva, N. F., Hanazaki, N., Albuquerque, U. P., Campos, J. L. A., Feitosa, I. S., & Araújo, E. D. L. (2019). Local knowledge and conservation priorities of medicinal plants near a protected area in Brazil. *Evidence-Based Complementary and Alternative Medicine, 2019*, 8275084.

DfID. (2021). *Annual Review*. DfID. Available at: https://devtracker.fcdo.gov.uk/projects/GBGOV-1-300461/documents. (last accessed 3rd May 2022).

Dintwa, K. F., Letamo, G., & Navaneetham, K. (2022). Vulnerability perception, quality of life, and indigenous knowledge: A qualitative study of the population of Ngamiland West District, Botswana. *International Journal of Disaster Risk Reduction, 70*, 102788.

Douglas, I. (2017). Flooding in African cities, scales of causes, teleconnections, risks, vulnerability and impacts. *International Journal of Disaster Risk Reduction, 26*, 34–42.

Eguaroje, O., Alaga, T., Ogbole, J., Omolere, S., Alwadood, J., Kolawole, I., et al. (2015). Flood vulnerability assessment of Ibadan City, Oyo State, Nigeria. *World Environment, 5*, 149–159.

Elagib, N. A., Al Zayed, I. S., Saad, S. A. G., Mahmood, M. I., Basheer, M., & Fink, A. H. (2021). Debilitating floods in the Sahel are becoming frequent. *Journal of Hydrology, 599*, 126362.

Enock, C. M. (2013). Indigenous knowledge systems and modern weather forecasting: Exploring the linkages. *Journal of Agriculture and Sustainability, 2*(1), 98–141.

Feka, N. Z., & Ajonina, G. N. (2011). Drivers causing decline of mangrove in West-Central Africa: A review. *International Journal of Biodiversity Science, Ecosystem Services & Management, 7*(3), 217–230.

Ford, J. D., King, N., Galappaththi, E. K., Pearce, T., McDowell, G., & Harper, S. L. (2020). The resilience of indigenous peoples to environmental change. *One Earth, 2*(6), 532–543.

Fuldauer, L. I., Thacker, S., Haggis, R. A., Fuso-Nerini, F., Nicholls, R. J., & Hall, J. W. (2022). Targeting climate adaptation to safeguard and advance the Sustainable Development Goals. *Nature Communications, 13*(1), 1–15.

Fuso Nerini, F., Sovacool, B., Hughes, N., Cozzi, L., Cosgrave, E., Howells, M., Tavoni, M., Tomei, J., Zerriffi, H., & Milligan, B. (2019). Connecting climate action with other Sustainable Development Goals. *Nature Sustainability, 2*(8), 674–680.

Gardner, J. S., & Dekens, J. (2007). Mountain hazards and the resilience of social–ecological systems: Lessons learned in India and Canada. *Natural Hazards, 41*(2), 317–336.

Garnett, S. T., Burgess, N. D., Fa, J. E., Fernández-Llamazares, Á., Molnár, Z., Robinson, C. J., Watson, J. E. M., Zander, K. K., Austin, B., Brondizio, E. S., Collier, N. F., Duncan, T., Ellis, Er., Geyle, H., Jackson, M. V., Jonas, H., Malmer, P., McGowan, B., Sivongxay, A., & Leiper, I. (2018). A spatial overview of the global importance of Indigenous lands for conservation. *Nature Sustainability, 1*(7), 369–374.

Hooli, L. J. (2016). Resilience of the poorest: Coping strategies and indigenous knowledge of living with the floods in Northern Namibia. *Regional Environmental Change, 16*(3), 695–707.

Iloka, N. G. (2016). Indigenous knowledge for disaster risk reduction: An African perspective. *Jàmbá: Journal of Disaster Risk Studies, 8*(1), 1–7.

IPBES. (2020). *IPBES data management policy ver. 1.0. Task force on knowledge and data* (R. M. Krug, B. Omare, & A. Niamir, Eds.). IPBES Secretariat. https://doi.org/10.5281/zenodo.3551079

IPCC. (2021). *Summary for Policymakers.* (V. Masson-Delmotte et al., Eds.). Climate Change 2021: The Physical Science Basis. Contribution of Working Group I to the Sixth Assessment Report of the Intergovernmental Panel on Climate Change IPCC.

Iwama, A. Y., Araos, F., Anbleyth-Evans, J., Marchezini, V., Ruiz-Luna, A., Ther-Ríos, F., Bacigalupe, G., & Perkins, P. E. (2021). Multiple knowledge systems and participatory actions in slow-onset effects of climate change: Insights and perspectives in Latin America and the Caribbean. *Current Opinion in Environmental Sustainability, 50*, 31–42.

Kalanda-Joshua, M., Ngongondo, C., Chipeta, L., & Mpembeka, F. (2011). Integrating indigenous knowledge with conventional science: Enhancing localised climate and weather forecasts in Nessa, Mulanje, Malawi. *Physics and Chemistry of the Earth, Parts A/B/C, 36*(14–15), 996–1003.

Karley, N. K. (2009). Flooding and physical planning in urban areas in West Africa: Situational analysis of Accra, Ghana. *Theoretical and Empirical Researches in Urban Management, 4*(13), 25–41.

Kasei, R. A., Kalanda-Joshua, M. D., & Benefor, D. T. (2019). Rapid urbanisation and implications for indigenous knowledge in early warning on flood risk in African cities. *Journal of the British Academy, 7*(s2), 183–214.

Korovulavula, I., Nunn, P. D., Kumar, R., & Fong, T. (2020). Peripherality as key to understanding opportunities and needs for effective and sustainable climate-change adaptation: A case study from Viti Levu Island, Fiji. *Climate and Development, 12*(10), 888–898.

Kurnio, H., Fekete, A., Naz, F., Norf, C., & Jüpner, R. (2021). Resilience learning and indigenous knowledge of earthquake risk in Indonesia. *International Journal of Disaster Risk Reduction, 62*, 102423.

Leal Filho, W., Barbir, J., Gwenzi, J., Ayal, D., Simpson, N. P., Adeleke, L., Tilahun, B., Chirisa, I., Gbedemah, S. F., Nzengya, D. M., Sharifi, A., Theodory, T., & Yaffa, S. (2022). The role of indigenous knowledge in climate change adaptation in Africa. *Environmental Science & Policy, 136*, 250–260.

MacLeod, D. A., Dankers, R., Graham, R., Guigma, K., Jenkins, L., Todd, M. C., Kiptum, A., Kilavi, M., Njogu, A., & Mwangi, E. (2021). Drivers and subseasonal predictability of heavy rainfall in equatorial East Africa and relationship with flood risk. *Journal of Hydrometeorology, 22*(4), 887–903.

Madhanagopal, D., & Pattanaik, S. (2020). Exploring fishermen's local knowledge and perceptions in the face of climate change: The case of coastal Tamil Nadu, India. *Environment, Development and Sustainability, 22*(4), 3461–3489.

Mannakkara, S., & Wilkinson, S. J. (2015). Supporting post-disaster social recovery to build back better. *International Journal of Disaster Resilience in the Built Environment, 6*, 126–139.

Manyena, S. B. (2014). Disaster resilience: A question of 'multiple faces' and 'multiple spaces'? *International Journal of Disaster Risk Reduction, 8*, 1–9.

Mapfumo, P., Mtambanengwe, F., & Chikowo, R. (2016). Building on indigenous knowledge to strengthen the capacity of smallholder farming communities to adapt to climate change and variability in southern Africa. *Climate and Development, 8*(1), 72–82.

Mavhura, E., Manyena, S. B., Collins, A. E., & Manatsa, D. (2013). Indigenous knowledge, coping strategies and resilience to floods in Muzarabani, Zimbabwe. *International Journal of Disaster Risk Reduction, 5*, 38–48.

McNamara, K. E., & Prasad, S. S. (2014). Coping with extreme weather: Communities in Fiji and Vanuatu share their experiences and knowledge. *Climatic Change, 123*(2), 121–132.

Murdoch, A., Mantyka-Pringle, C., & Sharma, S. (2020). The interactive effects of climate change and land use on boreal stream fish communities. *Science of the Total Environment, 700*, 134518.

Mutambisi, T., Chanza, N., Matamanda, A. R., Ncube, R., &Chirisa, I. (2020). Climate change adaptation in Southern Africa: Universalistic science or indigenous knowledge or hybrid. In *African handbook of climate change adaptation* (pp. 1–16). Springer.

Nakashima, D., & Krupnik, I. (Eds.). (2018). *Indigenous knowledge for climate change assessment and adaptation* (Vol. 2). Cambridge University Press.

Nyong, A., Adesina, F., & Osman Elasha, B. (2007). The value of indigenous knowledge in climate change mitigation and adaptation strategies in the African Sahel. *Mitigation and Adaptation Strategies for Global Change, 12*, 787–797.

Obi, R., Nwachukwu, M. U., Okeke, D. C., & Jiburum, U. (2021). Indigenous flood control and management knowledge and flood disaster risk reduction in Nigeria's coastal communities: An empirical analysis. *International Journal of Disaster Risk Reduction, 55*, 102079.

O'Donoghue, S., Lehmann, M., Major, D., Major-Ex, G., Sutherland, C., Motau, A., Haddaden, N., Kibria, A. S., Costanza, R., Groves, C., Behie, A., & Johnson, K. (2021). Adaptation to climate change in small coastal cities: The influence of development status on adaptation response. *Ocean & Coastal Management, 211*, 105788.

Okunola, O. H. (2022). *How policy can address frequent flooding in African coastal cities*. Africa at LSE.

Okunola, O. H., & Olawuni, P. O. (2022). Determinants of household flood resilience strategies in Kaduna Metropolis, Nigeria. *Urban Climate, 44*, 101216.

Olajuyigbe, A. E., Rotowa, O. O., & Durojaye, E. (2012). An assessment of flood hazard in Nigeria: The case of mile 12, Lagos. *Mediterranean Journal of Social Sciences, 3*(2), 367–367.

Patel, S. K., Mathew, B., Nanda, A., Mohanty, B., & Saggurti, N. (2020). Voices of rural people: Community-level assessment of effects and resilience to natural disasters in Odisha, India. *International Journal of Population Studies, 6*(1), 3–15.

Petzold, J., Andrews, N., Ford, J. D., Hedemann, C., & Postigo, J. C. (2020). Indigenous knowledge on climate change adaptation: A global evidence map of academic literature. *Environmental Research Letters, 15*(11), 113007.

Rahman, A., Sakurai, A., & Munadi, K. (2017). Indigenous knowledge management to enhance community resilience to tsunami risk: Lessons learned from Smong traditions in Simeulue island, Indonesia. *IOP Conference Series: Earth and Environmental Science, 56*(1), 012018.

Rentschler, J., & Salhab, M. (2020). *People in harm's way: Flood exposure and poverty in 189 countries* (Policy Research Working Paper: No. 9447). World Bank. License: CC BY 3.0 IGO. © World Bank. https://openknowledge.worldbank.org/handle/10986/34655

Salami, R. O., Giggins, H., & Von Meding, J. K. (2017). Urban settlements' vulnerability to flood risks in African cities: A conceptual framework. *Jàmbá: Journal of Disaster Risk Studies, 9*(1), 1–9.

Savo, V., Lepofsky, D., Benner, J. P., Kohfeld, K. E., Bailey, J., & Lertzman, K. (2016). Observations of climate change among subsistence-oriented communities around the world. *Nature Climate Change, 6*(5), 462–473.

Sayers, P., Yuanyuan, L., Galloway, G., Penning-Rowsell, E., Fuxin, S., Kang, W., Yiwei, C., & Le Quesne, T. (2013). *Flood risk management: A strategic approach*. Asian Development Bank.

Selemani, I. S. (2020). Indigenous knowledge and rangelands' biodiversity conservation in Tanzania: Success and failure. *Biodiversity and Conservation, 29*(14), 3863–3876.
Talubo, J. P., Morse, S., & Saroj, D. (2022). Whose resilience matters? A socio-ecological systems approach to defining and assessing disaster resilience for small islands. *Environmental Challenges, 7*, 100511.
Tengö, M., Brondizio, E. S., Elmqvist, T., Malmer, P., & Spierenburg, M. (2014). Connecting diverse knowledge systems for enhanced ecosystem governance: The multiple evidence base approach. *Ambio, 43*(5), 579–591.
United Nations. (2018). Revision of World Urbanization Prospects. *Population Division of the United Nations Department of Economic and Social Affairs (UN DESA).* https://esa.un.org/unpd/wup/
World Bank. (2017). *Implementing nature based flood protection: Principles and implementation guidance.*
Zhenmin, L., & Espinosa, P. (2019). Tackling climate change to accelerate sustainable development. *Nature Climate Change, 9*(7), 494–496.

National Resilience Leading to Sustainability; Case Study of Pakistan

Nadeem Ahmad Shah, Naeem Shahzad, and Syeda Uzma

1 Introduction

The Latin word "resilire" means to spring up; the word resilience came from the word resilire. This is defined as the ability of a nation/community to come back/spring back to the original condition or even better than the previous level. Resilience is such a unique character that if it is present in society will flourish very fast after every traumatic condition and the people of the community will perform in a much better way. This character is like the presence of antibodies that fight against the invading germs in the human body (Ilyas & Malik, 2016).

The term "resilience" is used for the people, the community, or a nation that has coped very well against the disastrous situation and bounced back with full strength. In order to find the resilience of a nation, it is imperative that, firstly we must know what actually resilience is. It is the ability to prepare for the forthcoming event and plan for the countermeasures, absorb the effects of the hazard, recover from the traumatic condition, and more successfully adapt to adverse events. A wide range of natural and man-made disasters comes under "Adverse events", and it is vital to distinguish that the resilience of a community against different types of disasters can be augmented by effectively planning for one kind of disaster in that locality (Disaster Resilience, 2012).

There are many factors on which the resilience of a community depends. These factors include:

- Access to the justice system of the community
- Physical conditions of the area

N. A. Shah · N. Shahzad (✉) · N. Shahzad (✉)
National University of Sciences and Technology, Islamabad, Pakistan
e-mail: naeemshahzad@mce.nust.edu.pk

S. Uzma
CMH, Risalpur, Pakistan

N. Shahzad (ed.), *Water and Environment for Sustainability*,
https://doi.org/10.1007/978-3-031-27280-6_11

- Individual capacities
- Population's social and economic conditions.
- Religious beliefs, bonding with religion
- Culture of the area
- Social bindings
- Level of awareness and education system of the community
- Level of cooperation

As it is variable, thus every community has different degrees of resilience, i.e., someone's are more resilient and others are not (Ahsen, 2017).

2 What Makes Pakistan Resilient?

Over the last many years, Pakistan has seen the brutal face of terrorism, faced extremism, sub-nationalist insurgencies, cross-border terrorism, conspiracies from the neighboring countries, many other problems like natural disasters, and other community issues. Resilience has got the power to effectively engage the social scientists of the world. It is such a unique capability of a nation that makes it able to come back to its original or even better condition after a disaster/calamity. Resilience is found in all the societies of the world and the community responds according to its strength and the available resources. Despite all the odds Pakistan has been facing, there is an innate strength of the Pakistani nation that makes it capable of withstanding all the different kinds of disasters and it distinguishes Pakistan from other nations. After Second World War, Pakistan is the only country that suffered a vicious disintegration within 24 years of its existence. Pakistan suffered a violent disintegration after Second World War. Since the creation of Bangladesh, Pakistan has existed as a compact state for more than four decades, despite vicious conspiracies by hostile countries. Islam is the main religion of the majority of the country which offers choices and autonomy to the followers while practicing the religion. It is a state religion in the country but its practice is a privatized affair. It may be attractive to discover some of the fundamentals that make Pakistani society resilient. When we tolerate the impact of adverse events, there is also impulsive healing happening across all community layers and this is because of the closely inter-woven social texture of the country (Mahmud, 2015).

3 Originality/Value

Raising the importance of resilience for a nation is the original part of this paper, especially for developing countries to achieve a post-disaster resilient environment and to minimize the losses of lives and traumatic situations. The paper also offers a vision to continue the normal functioning of everyday life even after the disaster situation.

4 International Perspective

Resilience is a process and not a personality characteristic. It mainly arises when some nations interact with risk and protective factors. Most of the time resilience is seen as an outcome, but this is not good. It is a response to difficult circumstances when a nation or a community comes across these kinds of disastrous situations. It is not something people have or do not have, but it is based on the way people respond at that time. There are different types of resilience; however, depending upon the situation of the individual or a community, some kinds of resilience may be considered healthier than others. Resilience changes with time as it is not static. Because of the presence of protective factors, resilience may appear at one stage of disaster but disappear at some other stage. As people may be facing different sets of positive and negative circumstances, so we cannot blame people if they are seen to be less resilient than others (Coleman, 2007).

To prevent conflict and to move away from fragility toward resilient nations, States increasingly adopt institutionalized mechanisms and systemic efforts. However, under these situations, it is important to note that these efforts must be supported by United Nations Development Program (UNDP), and partners at the national level and examines how they have contributed to transforming conflict situation and improving national governance (Ryan, 2012). Efforts need to be channelized around legislation and policy, implementation of resources, development of industry and market, and resourcing access and transportation system to have a resilient and sustainable built environment after a disaster (Yan Chang et al., 2010).

The capacity for successful adaptation in the face of disturbance, stress, or adversity, found by research in social sciences and biological sciences, is indicated by resilience. After a disaster event, broadly speaking, there are five dimensions to resilience which include socio-psychological well-being at social and family levels, restoration at the organizational and institutional level, the commercial and economic continuation of services and productivity, reinstating infrastructure, and regularizing public safety (Aldrich, 2012). Disasters result in the loss of human lives, damage to infrastructure, and loss of economic activity in the community. In order to rebuild a community and foster resilience, the role of women is considered crucial. Building resilience requires more than just reducing vulnerability. The aim of enhancing resilience is to enhance the ability of local people to respond effectively against a disaster. In order to build the resilience of a vulnerable group, a strong community and the assistance of government departments are required. These departments and community will help to support efforts to cope with the negative effects of disaster and to support the vulnerable group of the community, offering social protection and development initiatives. Under the UN International strategy for disaster reduction, governments need to find a new development in the form of linking community-based disaster risk reduction (CBDRR) with sustainable development goals. While ensuring the resilience capacity of communities and DRR measures that are considered very important for development, there is always an opportunity to 'build back better', after a disaster (Drolet et al., 2015). About Pakistan's demise, rumors have been greatly exaggerated. The country defied the odds, and by any measure is one of the most resilient nations on earth. Pakistan has been periodically subjected to survive man-made and natural catastrophes and has got an inherent ability to rise from the ashes (Sehgal, 2010).

5 National Perspective

We are so much more than what the world sees if we stand united. Although we are facing sectarianism, bribery, corruption, poverty, and illiteracy, many other things show the bright face of the nation. We are also facing terrorism, violence, guns, and bombs; however, there is more to this country that shows the level of resilience of this nation. We are the essence of flexibility (Sheikh, 2014). Pakistan is a strong and resilient nation that has withstood many upheavals and crises with great composure. Those who say that this is a lost nation are sadly mistaken. This nation has had to struggle but in no way is it a failed state. There are many challenges, but the people of Pakistan are more than willing and able to deal with them (Yaseen, 2011). Pakistan has suffered immensely for the last few decades because of terrorism. Throughout the history of Pakistan, we have been liberal, progressive, and forward looking. Despite the hard times, Pakistan has recently been in the process to hand over the civilian government to another democratically elected government. This is the evidence of our commitment and perseverance in the face of chaos, tensions, and hostility. This is what we call resilience. Chris Jaffrelot in his recent book called that Pakistan not only survived but also has the tendency to thrive. Pakistan has been the only country in the world to suppress a full ranging insurgency in Swat in the twenty-first century. Pakistan cannot be compared with Iraq and Afghanistan as it has a very strong military, backed by the people of Pakistan (Farooq, 2017).

If we ask any Pakistani, in any part of the country, he will tell us the hardships Pakistan has been facing and even then we as a nation remained united and restored effectively to bring the nation to the routine activities, without hampering the morale of the people of Pakistan. Even our most bitter critics will pay honor to our resilience and firm capacity for withstanding all tests. To be Pakistani means being able to adapt to any situation, and this has become the keystone of our uniqueness. It is difficult or impossible to identify priorities for improvement, determine whether resilience has improved or worsened, or compare the benefits of resilience with the associated costs, without a good measure of resilience (Ahmad, 2015). There is a long list of disasters; Pakistani nations coped successfully; however, a few significant events are discussed here to find the level of resilience of the nation.

6 Separation of East Wing of Pakistan (1971)

With the creation of Pakistan and India as two independent states, British rule in the Indian subcontinent came to an end in August 1947. In the eastern and western sides of India, the state of Pakistan comprised of the Muslim-dominated areas, separated by thousands of miles of Indian territory. These two areas were named "East Pakistan and West Pakistan". Soon after independence, to run the affairs of the state, Pakistan had no pre-established administrative site or a sizable number of skilled bureaucrats (Nuruzzaman, 2010). The separation of the East wing of Pakistan is considered the most tragic moment in 73 years history of Pakistan. In 1971, Pakistan faced not only the separation of its East wing but also a civil war in the country and the 3rd bloody war with India that lead to an unending series of miseries. There is a long list of factors that contributed to the great tragedy; however, sense of dispossession within East Pakistan was one of the main reasons. In addition to it, the demand for provincial autonomy and a greater share of the resources of the country is also on the list (Khan, 2011). To identify the underlying causes of this tragedy, Hamood-ur-Rehman commission was constituted by the Government of Pakistan. Among other causes, the failure of the then military and political leadership in resolving the crisis was described as the main reason by the commission in its report. The exaggerated figures for the killings of Bengalis by the Pakistan Army were rejected by many analysts, investigating journalists and writers (Jamil, 2017). After disintegration, the Pakistani nation was depressed and felt the pain of separation, just after 24 years of independence and blamed the military and the political elite of that time, to be responsible for that disaster. But the world witnessed that Pakistan survived and is progressing even after four decades after that sad event (Fig. 1).

7 Ojhri Camp Disaster (1988)

The Pakistani nation will remember April 10, 1988, as a day of mass grief. The twin cities of Islamabad and Rawalpindi received a horrible shock on that day, early in the morning. Initially, there were explosions of low density, followed by a huge explosion. Missiles, rockets, and projectiles started raining down in all directions. The people of Rawalpindi and Islamabad were in a lot of panic and fear. There were rumors ranging from an Indian attack to a man-made disaster at some location in the city. No one could explain what had happened. Later on, the government officials started informing the local people that there was just an accidental explosion at the arms and ammunition depot at Ojhri, located between Rawalpindi and Islamabad, and there is no reason for fear. Because of this incident in the populated cities of Islamabad and Rawalpindi, over 100 people were killed and many more were injured. However, after such a disastrous situation where people experienced the taste of actual war with bombs and rockets coming from all directions, the daily routine of life started the very next day (Aziz, 2016).

Fig. 1 Location of East and West Pakistan after partition (Singh, 2017)

8 Earthquake (2005)

More than 70,000 people were killed and 3.5 million were rendered homelessness after a devastating earthquake jolted the northern part of Pakistan on the 8th of October, 2005. After this disaster, the Pakistani nation rose with such an unmatched demonstration of commitment and concern that never happened in the history of Pakistan. Soon after the disaster, the response was purely humanitarian with less emphasis on organizational matters. In the difficult and mountainous terrain of the affected area, Pakistan's armed forces worked with great commitment to provide relief and rescue services. Pakistan army and other civil rescue organizations had to face difficulty in access to the disaster area because of landslides, road blockage, and damaged bridges. In the past, Pakistan has never been hit by such a gigantic disaster of a high degree. To provide relief goods from all parts of the country to the disaster area and to provide facilities to those in urgent need, the government and the people of Pakistan joined hands together. Welfare Trusts like Edhi, faith-based organizations like Pakistan Islamic Medical Association (PIMA), Alkhidmat, etc. worked round the clock. The efforts of all these organizations and institutes were exemplary. Devotion to service is a lesson that Islam as a religion teaches to humanity. Relief supplies were transported by the international community through helicopters. To help the victims and to transfer seriously injured people to the hospitals, volunteers and rescue teams from eighty countries helped Pakistani authorities. After the earthquake, one thing

Fig. 2 Earthquake 2005 in Pakistan (Artibees, 2018)

that came out with a strong message that the Pakistani nation not only came out to help the sufferers but showed an example of devotion, dedication, humbleness, and generosity toward the victims of that calamity. Even after the main quake, tremors were continually felt in the region, but the spirit of the people remained unshakable. By no means could the disaster break the spirit of the people, although there were so many people who had lost their relatives and friends, or homes, or had been injured or had lost their body parts due to trauma. The sense of thankfulness and gratefulness to God was awe-inspiring even after trauma and miseries. During those testing times, the people of Pakistan took shelter in their religious beliefs, asking God for assistance. The people were of the view that Allah gives and takes. Among the victims of the earthquake, all that might have come because of belief in prayers and faith in religion. This shows that the power of resilience is vital for the survival of any nation. It is also crucial that people bounce back and establish themselves as they were prior to the disaster. There is no example of such strength of faith and spirit. By a sudden shake of the earth, thousands were dead, wounded, and displaced. At one moment one feels gloomy and depressed over the incident; however, there is always a hope to recover with time (Niaz, 2006) (Fig. 2).

9 Devastating Floods (2010)

In July 2010, an easterly monsoon system collided with the western system in Pakistan and a weather system developed over North-Western Khyber Pakhtunkhwa (KP) province that brought along with it the destruction and devastation of unprecedented scale, in the recorded history of Pakistan. This heavy rain and flooding that started in the Swat and Kabul Rivers in Khyber Pakhtunkhwa (KP) province shifted downstream to bring havoc to the life and property in other parts of the country. The data was compiled by the federal flood commission (FFC), and the magnitude of the calamity was gauged using that data. The worst flood recorded in the history of Pakistan before 2010 happened in 1929 when the flow of rivers was recorded at 250,000 cusecs. However, in 2010 flooding, the flow of rivers Swat and Kabul combined touched a new historical height of 400,000 cusecs. The flood waters traveled downstream through the barrages in Punjab and Sindh provinces, after playing havoc with life and property in Khyber Pakhtunkhwa province, until it reached the Arabian Sea. As a result of this massive flow of water, there was a long track of death and destruction from north to south of the country. There were 1980 reported deaths, and nearly, 2,946 were injured because of the calamity. As a result of this disaster, more than 20 million people in Pakistan were affected. Around 1.6 million homes were destroyed, and thousands of acres of crops were damaged. The public infrastructure also sustained damages at an enormous level. A total of 25,088 km of roads, 10,436 education centers, and 515 health facilities were damaged. On account of direct and indirect damages to social and physical infrastructure, economic sectors, governance, and environment, Pakistan had to face an estimated loss of USD 10 billion, reported by the World Bank (WB) and Asian Development Bank (ADB). In short, the scale of devastation caused by the worst ever catastrophe was more than the Pakistan Earthquake, in 2005, Cyclone Katrina, the Indian Ocean Tsunami, Cyclone Nargis, and the Haiti Earthquake combined in terms of area and population (Zulfiqar, 2010) (Fig. 3).

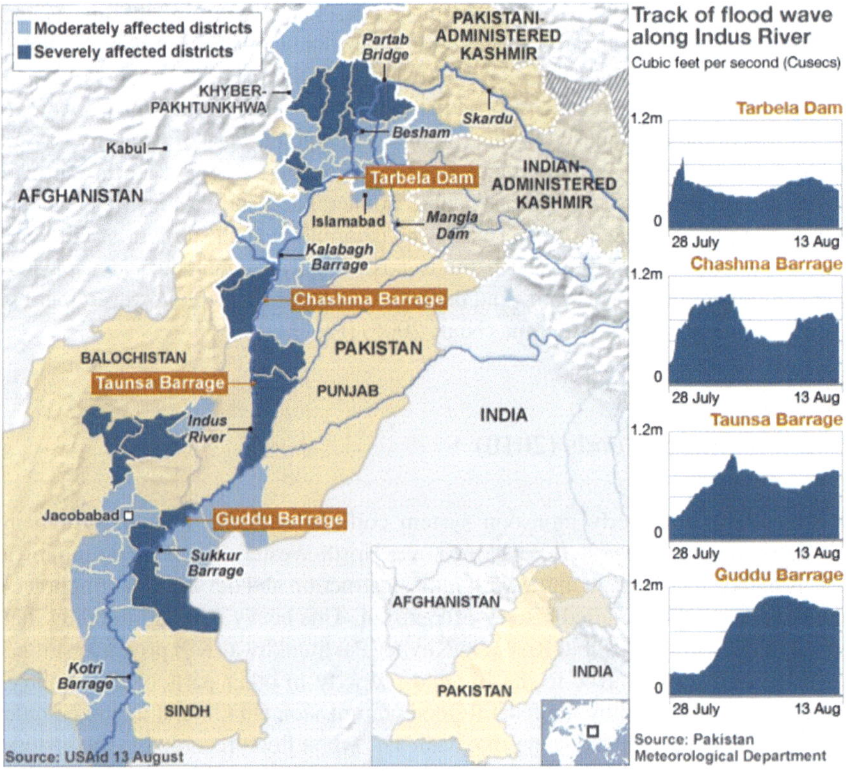

Fig. 3 Devastating floods in Pakistan (BBC, 2010)

10 Gayari Sector Siachen (2012)

An avalanche hit a military base at the Gayari sector in Siachen Glacier on April 7, 2012. As a result of this incident, 140 soldiers and civilian contractors were trapped under deep snow. It was the worst tragedy that the Pakistani military has experienced in the area. The incident occurred around 180 miles North East of Skardu, at an altitude of 16,000 feet (Shaukat, 2015). The soldiers of the Pakistan Army were buried under 70 feet of snow after this disaster in the region (Abbas, 2016). The soldiers were buried under a boulder with dimensions of 120 foot long and 70 foot wide. The army, after the incident, started all possible efforts to search for all its soldiers buried alive under the boulder. After continuous efforts, Pakistan Army and other agencies recovered all the bodies and handed them over to the heirs. It was thought to be an impossible task to find the bodies of all those buried under the ice cover at such high altitudes under unpredictable weather conditions. International experts were of the view that it is a useless effort to find bodies under thick snow cover and suggested declaring the site a mass grave. There was a consensus among a few other experts who even said that seven to eight years would be needed to dig out the bodies. The army was firm in its determination and decided to put in its best efforts and resources, despite all the difficulties. It was the unprecedented courage and persistence of the jawans of the Pakistan Army that made it possible to complete the task within one year. After working day and night, the bodies of martyrs were recovered with the help of excavation equipment. The Frontier Works Organization (FWO), National Database and Registration Authority (NADRA), the international community, and other organizations played a key role in helping Pakistan Army in search operations (*Hussain*, 2012) (Figs. 4 and 5).

Fig. 4 Gayari sector tragedy (Alchetron, 2012)

Fig. 5 Gayari sector military base (Tribune, 2017)

11 War on Terror (2001 Onwards)

The global war against terrorism started after terrorist attacks on the United States in 2001. During the last almost two decades, Pakistan has been a victim of instability and insecurity due to its role as the main state in this combat against terrorism. From 2001 to 2017, more than seventy two thousand people have been killed in terrorist attacks in Pakistan. After this war, Pakistan is facing security challenges and has paid the high cost concerning its financial system, cost of investment inflows, market confidence, and social fabric deformation (Ahsen, 2017).

12 The Assassination of Benazir Bhutto (2007)

Benazir Bhutto, the former Prime Minister of Pakistan, was the first woman to lead a Muslim country. A gunman shot her in the neck and set off a bomb on 27 December 2007 when she was leaving an election rally in Rawalpindi. As a result of this attack, at least 20 other people died and several more were injured (Jones, 2007). She was the most popular leader of the country and was also an active Chairperson of the largest political party in Pakistan. The killing of Benazir Bhutto was carried out with the aim to destabilize Pakistan. People of Pakistan reacted strongly across the country and again the nation showed its resilience and stood firm against the attack and showed solidarity with the families of all those martyred in that attack (Fig. 6).

13 Attack on the Sri Lankan Cricket Team in Lahore (2009)

A bus carrying the Sri Lankan cricket team to Lahore's Qaddafi Stadium for Day 3 of their 2nd test match against Pakistan was targeted by militants on March 3, 2009. This terrorist attack was carried out on the game which is arguably the most celebrated game in the country. Cricket fans across Pakistan regret the awful attack,

Fig. 6 Benazir Bhutto in her last rally in Rawalpindi (Taj, 2010)

especially when the attack was carried out on foreign cricketers. At around 8:50 am, terrorists stormed through Liberty market, Lahore intending to carry out a multi-pronged attack on the Sri Lankan team's convoy. Using RPGs, hand grenades, and guns, the attackers aimed the bus carrying the cricketers. Six police officers in an escort van were killed, and six cricketers of the Sri Lankan cricket team were injured. Two bystanders were also killed. The Lahore test match was quickly called off and the tour was canceled, with the Sri Lankan players evacuated from the Gaddafi stadium and taken to a nearby airbase. Because of the deteriorating security situation in the country, most of the cricket teams had already regretted visiting Pakistan. This incident was heartbreaking not only for the Pakistani nation but also for all cricket lovers all around the world. The cricketing nations boycotted Pakistan for playing cricket in the country. Pakistani stadiums were turned into deserts, and the situation remained the same for around a decade. Pakistan was forced to play the scheduled home matches in Dubai and Sri Lanka (Chaudhry, 2017). The resilience and resolve of the Pakistani nation worked here as well. Even after not being able to play cricket at home grounds, Pakistan managed to win the ICC Champions Trophy 2017 final against archrival India, hosted by England (Alter, 2017).

14 Attack on Army Public School (APS) Peshawar (2014)

In December 2014, terrorists attacked Army Public School (APS) and killed dozens of students and teachers. In our recent history, this massacre may be considered a turning point, when people from all the different parts of the country came together and made a strong bond against terrorism. All institutions, politicians, media, and civil society of Pakistan came on one page against all forms and manifestations of terrorism. After the Peshawar tragedy, under a cohesive, well-orchestrated National Action Plan (NAP), the back of terror has been broken very effectively by law enforcement agencies. Unfortunately, it took a while; however, the resilience of the nation has paid back and the results are there to witness as terror incidences are on a rapid decline (Sharif, 2015).

15 Attacks on Military Bases

Minhas air base of Pakistan Air Force came under attack on August 16, 2012, when militants carried out a blatant attack while taking advantage of darkness. During a gun battle that raged for hours after the raid, eight militants were killed while one soldier was also martyred. One aircraft was damaged, and the attack was repelled successfully (Malik, 2012). Four terrorists launched an assault with sophisticated weapons on Pakistan Army Headquarters (GHQ) on October 10, 2009. Six soldiers including two senior army officials were martyred in a noxious terrorist attack. Militants took at least 42 hostages, and the GHQ remained under blockade for several hours. Pakistan Army rescued 39 hostages after operation Janbaz was launched; however, during the attack and operation, 11 soldiers and three hostages were killed (Rao, 2009). One of Pakistan's naval bases was attacked by militants on May 22, 2011. Around 15 terrorists stormed into the base, carrying guns, grenades, and rocket-propelled grenades. An aircraft, used for surveillance, and a helicopter were destroyed by militants (Waraich, 2011).

16 The Resolve and Commitment of the Pakistani Military

Pakistani military, with Pakistan police, rangers, and other Law Enforcement Agencies (LEAs) remained under the continuous attack of the militants in all parts of the country. Initially, the Law Enforcement Agencies (LEAs) reacted defensively but with the passage of time, these militants started attacking the nerve areas of Pakistan's military forces. It was high time for the morale of the Pakistani military when all the nodal points of military setups were attacked by the militants regularly and no offensive response was initiated in reaction to these embarrassing attacks. Pakistan's armed forces were criticized for not being offensive against the terrorists. Here at this point, the Pakistan army started an offensive campaign, soon after General Raheel Sharif took over command as Chief of the Army Staff. Pakistan Air Force was also effectively utilized in order to provide aerial cover and to destroy the nerve points of the militants with aerial bombardment, all along the western border. Pakistan's military showed full commitment and resolve and started military operations one after the other, eliminating the terrorist which resulted in a substantial decrease in the terrorist attacks on the civil and military setups all around the country. This action was highly appreciated by world leaders and international media. The world started to recognize the efforts carried out by the military and the sacrifices of the people of Pakistan.

17 Military Operations in Pakistan

Apart from tension at the border with India and Afghanistan, Pakistan has been engaged in four wars with neighboring countries. For a long time, the Pakistan army has had to battle the terrorists within the country. Pakistan joined hands with the US-led War on Terror and helped.

The US forces were involved in the process of eliminating the terrorists, after the September 11 attacks in the United States of America. All the military operations conducted by the Pakistan military against terrorists have been successful. In recent years, the Pakistan military has undertaken many joint operations (Table 1); however, in eliminating terrorists, Operation Rah-e-Rast and Zarb-e-Azb have been the most successful. Operation Zarb-e-Azb is the biggest and most well-coordinated operation ever conducted against terrorists. As it is a 'war of survival', hence this operation holds greater significance among all the operations conducted so far (Nabi, 2016).

Table 1 Military operations in Pakistan (Nabi, 2016)

No	Military operations	Year	Remarks
1	Al-Mizan Operation	2002	• First operation at start of war against terror • Around 70,000–80,000 men were deployed • Special Services Group also participated
2	Operation Rah-e-Haq	2007	• Offensive but poorly coordinated • Operation conducted after Lal Masjid Incident • Emergence of Tehrik-i-taliban Pakistan (TTP)
3	Operation Sherdil	2008	• Conducted in Bajaur agency against militants
4	Operation Zalzala	2008	• 0.2 million people displaced • Around 40,000 houses destroyed
5	Operation Sirat-e-Mustaqeem	2008	• Operation in "Bara Tehsil"
6	Operation Rah-e-Rast	2009	• Pak Army regained control over 'Mingora'
7	Operation Rah-e-Nijat	2009	• Use of gunship helicopters and aircrafts
8	Operation Koh-e Sufaid	2011	• Tal-Parachinar road was re-opened by military
9	Operation Zarb-e-Azb	2013	• Effective utilization of force with tactics • Pakistan Air Force was effectively utilized
10	Operation Radd-ul-Fasaad	2017	• A nationwide multiagency response • Participation of Military and civil armed forces

18 Conclusion

Pakistan has been facing so many odds since the time of her independence. It had to face war with India, within the first year of its birth in 1948, followed by wars in 1965 and 1971. In addition to the wars, Pakistan has been experiencing different forms of natural and man-made disasters in the shape of floods, earthquakes, landslides, droughts, epidemics and terrorism, etc. All these hazards have tested the resilience of the nation, and Pakistan has proved to be the most resilient nation in the world which is not only coping after the disaster but still flourishing and competing with the world in all the different fields. No other nation would have survived and flourished after the disastrous events Pakistan faced. Pakistan has got the natural healing capacity to cope with the disaster timely and effectively and one of the reasons is a strong belief and faith in the religion. Some events were highly disturbing (separation of the East wing of Pakistan, attack on the Army Public School, Peshawar, and a wave of terrorism across the country) and shattered the confidence and the morale of the nation. People all around the country faced the loss of their loved ones while fighting against terrorism. Because of the resilience of the nation, as a whole, Pakistan fought back strongly after every incident. The worst enemies of Pakistan also acknowledge the level of motivation and resilience of the Pakistan military and the nation against all these odds. The recovery of all bodies that were buried under the avalanche in Gayari sector in 2012 is one example showing the resolve and the determination of the whole nation.

References

Abbas, M. (2016). On this day four years ago: A Gayari avalanche strikes Pakistan Army base in Siachin. *The Express Tribune.*

Ahmad, T. A. (2015). Himmat na Harna: Is Paksitan's Resilience also its undoing, Article. *Dawn.* http://www.dawn.com/news/1224774

Ahsen, S. B. (2017). *Can community building help end extremist violence in Pakistan?* Global Village Space.

Alchetron. (2012). Gayari sector avalanche. *Alchetron.* alchetron.com/Gayari-sector-avalanche

Aldrich, D. P. (2012). *Building resilience.* The University of Chicago Press.

Alter, J. (2017). How Pakistan won the ICC Champions Trophy. *Times of India.*

Artibees. (2018). Earthquake 2005 in Pakistan, news clip. *Artibees.* artibees.com/2018/10/08/earthquake-2005-in-pakistan/

Aziz, S. (2016). The ojhri camp disaster-who is to blame? *Dawn.*

BBC. (2010). Pakistan floods; maps and graphics, news clip. *BBC South Asia.* bbc.com/news/world-south-asia-1098620

Chaudhry, H. (2017). 8 years ago today: Attack on the Sri Lankan cricket team. *Dawn.*

Coleman, J. (2007). *Adolescence, risk and resilience: Against the odds.* Wiley.

Disaster Resilience. (2012). *The National Academies.* The National Academy of Sciences.

Drolet, J., Dominelli, L., Alston, M., Ersing, R., Mathbor, G. M., & WU, H. (2015). women rebuilding lives post-disaster: Innovative community practices for building resilience and promoting sustainable development. *Gender and Development, 23*(3), 443–448.

Farooq, K. K. (2017). The inherent resilience of Pakistan. *Pakistan Defence*, 1–3.

Hussain, H. (2012). Snow temple. *Defence Journal, 15*(9), 55.
Ilyas, R., & Malik, F. (2016). Development of a scale of resilience against terrorism. *Pakistan Journal of Social and Clinical Psychology, 14*(1), 15–20.
Jamil, M. (2017). Facts and fiction—1971 events. *Pakistan Observer*.
Jones, O. B. (2007). *Benazir Bhutto assassination: How Pakistan covered up killing*. BBC.
Khan, I. A. (2011). *1971-Lessons Et Cetera*. Criterion Quarterly Magazine.
Mahmud, T. (2015). What makes Pakistan resilient? *The Express Tribune*.
Malik, Y. (2012). Terrorists attack Kamra Air Base, nine attackers dead, plane damaged. *Dawn*.
Nabi, A. (2016). *List of military operations in Pakistan*. Live Rostrum News Agency.
Niaz, U. (2006). Role of faith and resilience in recovery from psycho-trauma. *Pakistan Journal of Medical Sciences, 22*(2), 204.
Nuruzzaman, M. (2010). Federalism and state disintegration—United Pakistan, 1947–1971: Some historical lessons for Afghanistan and Iraq. *Journal of Asian and African Studies, 45*(5), 504–521.
Rao, S. (2009). Terror attack on GHQ. *The Nation*.
Ryan, J. (2012, December 1). Infrastructure for peace as a path to resilient societies: An institutional perspective. *Journal of Peacebuilding and Development, 7*(3), 14–24.
Sehgal, I. u.-M. (2010). *Pakistan: A Resilient Nation*. EastWest Institute.
Sharif, M. S. (2015). Pakistan: A story of hope and resilience. *The Nation*.
Shaukat, S. (2015). *Memory of the Siachen Glacier Tragedy*. zameer36.
Sheikh, I. (2014). Pakistan: The epitome of resilience. *The Express Tribune*.
Singh, K. (2017). Train to Pakistan. *Word by Word*.
Taj, A. (2010). Benazir's killers? Years three and counting. *Pakistan Today*. pakistantoday.com.pk/2010/12/27/benazirs-killers-year-three-and-counting/
Tribune. (2017). Five years on, Pakistan remembers Gayari tragedy. *The Express*. tribune.com.pk/story/1377085/five-years-later-pakistan-remembers-fateful-gayari-sector
Waraich, O. (2011). Pakistan extremists embarrass military by invading a Navy base. *Time*.
Yan Chang, S. W., Seville, E., & Potangaroa, R. (2010). Resourcing for a resilient post-disaster reconstruction envirnment. *Emerald Insight*.
Yaseen, N. (2011). A resilient nation. *The Pakistan Today*.
Zulfiqar, H. (2010, May–December). *NDMA website*. Retrieved from NDMA Newsletter: http://www.ndma.gov.pk/publications/NLMay-Dec2010.pdf

Community Post-flood Vulnerability Assessment of Tehsil Behrain District Swat, Pakistan

Shams Ul Haq and Naeem Shahzad

1 Background

Pakistan has witnessed 7 major floods since its creation. Among, the flood 2010 devastated the socioeconomic and environmental infrastructure of the entire country generally and specifically in District Swat. The occurrences of natural disasters are common phenomena due to geographic features and the presence of active fault lines. However, such a catastrophic flood has never been seen in the history of Pakistan. It destroyed buildings, business centers, markets and houses, washed away agricultural land and vegetation cover, bridges, cut down roads and destroyed important lifelines. It caused massive displacement and put constraints on food security, livelihood, education and health facilities. The massive damages and losses were due to the fragile vulnerability conditions of people with a lack of resilience to withstand flood disasters. Similarly, no mechanism of early warning was in place at any level to disseminate early warnings and facilitate safe evacuation.

Many studies including Pakistan flood 2010 by Asian Development Bank (ADB), District Swat Disaster Management Plan (DDMP) and Behrain Tehsil Disaster Management Plan (TDMP) by Provincial Disaster Management Authority (PDMA), underscored the scale of damages and losses from flood 2010. As a result, access to basic facilities and services and constrained livelihood options within the local communities increased people's vulnerability. The situation got worse in those areas of upper Swat where government and aid agencies were unable to get access to flood affected people and serve humanity.

S. U. Haq (✉)
Pakistan Red Crescent Society, Islamabad, Pakistan
e-mail: ad.cva@prcs.org.pk

N. Shahzad
National University of Sciences and Technology, Islamabad, Pakistan
e-mail: naeemshahzad@mce.nust.edu.pk

N. Shahzad (ed.), *Water and Environment for Sustainability*,
https://doi.org/10.1007/978-3-031-27280-6_12

In the aftermath of the emergency response, the Government of Pakistan-GoP, humanitarian, UN and aid agencies initiated several programs focusing on the "build back better" notion. The dislocated households returned to their dwellings. The market steadily got functional which gave a ray of hope for livelihood revival. However, disaster resilient development is essential for long-term sustainability backed by research-oriented policies and programs.

2 Overview of the Study Area

Tehsil Behrain is widely famous for its scenic natural beauty and moderate weather during the summer season. District Swat has four Tehsils (Sub-District or Administrative Units) and Behrain is one of the administrative Tehsils located in the north of the capital city Mingora of District Swat. Multiple streams are falling in the river Swat at the catchment area of Behrain. River Swat is flowing through the center of this Tehsil. It is located at a distance of 316 km from Islamabad. The total population is 248,474 (census 2017–2018). The Tehsil is surrounded by mountains from all sides. It receives an average rainfall of 450–700 mm/year.

The moderate weather condition in summer attracts tourist from all parts of the country including international tourists as well. The average high temperate reaches up to 21 °C in June and becomes harsh at 1 °C or −0 °C during winters in January. The area does not count for monsoon season therefore occasional rainfalls are witnessed in this season. The average rainfall is around 450–700 mm/year.

The limited livelihood opportunities, education and geographic features make it prone to multiple disaster effects. Flood 2010 notably affected the entire Tehsil Behrain as people lost assets, livelihood sources, stranding crops and infrastructure. The people in the area are mainly associated with agriculture, livestock and the tourism industry, however, both sectors have been substantially affected due to recurrent disasters in the area. People were practicing subsistence agriculture for survival; however, the flood 2010 washed away the productive lands, standing crops, and stored produce and damaged the overall agriculture infrastructure of the village (Bank, 2010). The irrigation channels in some of the communities were also damaged which resulted in a decrease in productivity (community assessment). Major crops grown are mostly cereals and vegetables, while apple, peach and persimmon are the major fruit orchards that have provided employment opportunities to a large number of people. The flood affected the livelihood of these people as they were not able to transport their fruits and vegetables to market for cash returns. The flood destroyed bridges and roads that restricted access to essential services, resulting in increased prices of staple food items.

The major tribes living in the village are Yousafzai, Pakhtuns, Gujars and Mians. The women use veils and generally are not allowed to go out freely from their house boundaries. People live in a joint family system and had strong social bonds before the flood disaster. The flood left many families in isolation, many became temporarily dislocated people-TDP and many lost their native village and land. The social bond and cohesion weakened over time. While addressing the humanitarian needs of different organizations in the early recovery and recovery phase, led the local people to compete to get more out of it as individuals rather than a community. However, some best practices for instance sharing of accommodation, agriculture and other required tools were also observed in the study area during this phase.

Though people have high affiliations with politics to follow personal interests and have no or little understanding of their roles and responsibilities. The political leadership was never held accountable and was held responsible for the failure to develop a progressive and resilient society. The benefits for disaster affected people were often politicized, with the focus on party supporters to gain loyalty in future elections.

3 Floods 2010 in Swat

The devastation caused by the flood 2010 in Pakistan will always be remembered. The flood 2010 affected more than 20 million people. The devastations of the flood were widespread but District Swat was particularly affected. The Upper District Swat is prone to multiple hazards including floods, earthquakes, heavy rainfall, snowfall and thunderstorm. High levels of poverty, high population densities, environmental degradation, lack of political will and lack of coordination, planning and preparedness make the region highly vulnerable to natural disasters. The geographic obstacles such as steep, high mountains and harsh weather conditions restrict access to these locations.

In the study area, disaster risk reduction (DRR) has not been mainstreamed into development which has resulted in huge devastation due to recurring disasters (natural and anthropogenic). This study aims to identify factors that increase people's vulnerabilities and reduce coping capacities. An indicator-based study on exposure to flood, access to facilities, existing mitigation measures, socioeconomic culture and weak relations has been evaluated for an in-depth understanding of vulnerabilities.

This study examines community vulnerability by vulnerability assessment index-VAI taking into account the three parameters including exposure, fragility and lack of resilience. The priority hazards identified in the consultation include major floods, flash floods and heavy rainfalls. These indicators were used to identify the community's exposure to floods. Similarly, income, education, health, social networking, house structure and environment were considered indicators for the identification of fragility and lack of resilience. Several underlying risk factors increase vulnerability and contribute to the severity of disasters in the study area. These include:

- Poor construction practices and limited enforcement of existing building codes
- Weak early warning systems
- Lack of awareness and education on disasters and response
- Limited capacity and coordination between various government disaster response agencies

- Disaster susceptibility of a large number of impoverished communities.

Limited attention has been given to analyzing the key reason for vulnerabilities causing widespread destruction at the time of disaster. The analysis and results of vulnerabilities assessment provide foundations for effective planning for disaster risk reduction and management. The insights from this study are likely to provide the basis for effective planning at the government level to combat disasters in the future.

4 Previous Studies Related to Vulnerability Assessments

4.1 International Studies

According to Morrow (1999) social and economic conditions determine disaster vulnerability. The study highlights that disasters restrict the resilience of a single-headed household or households having large numbers of dependents who most of the time juggle multiple jobs to meet family needs. The study explained that vulnerable people such as the poor, senior citizen and women-headed household suffer more from disasters. Mapping these groups within the communities is essential for effective response. The study also recommended the engagement of these groups in planning and response activities in order to promote local activism which will leave a long-lasting impact on disaster resilience.

Fekete used the social vulnerability index intending to develop and validate a social vulnerability map of people prone to river floods in Germany. The composite index was used with three indicators (fragility, socioeconomic condition and region) adopted from the office of federal statistics (Fekete, 2009). The households in the three federal states were surveyed and logistic regression was used for the analysis of the vulnerability. The results of the social vulnerability index presented showed varied vulnerability conditions of different counties in Germany.

Mwape (2009) conducted a study on flood impacts on the socioeconomic condition of people and livelihood in Zambia. The study carried out the qualitative and quantitative analysis. The critical elements including agriculture, health and education, access to water, assets/property, housing and sanitation were found to be drastically affected by the flood. The damages and losses to the critical aspects increased the socioeconomic vulnerability of the people.

According to Cutter et al. (2012), disasters have become more devastating than before. The reasons are the hazard's frequent interaction with the widespread distinct vulnerabilities in the US County. The hazards of place model were used in the study which concluded that vulnerability is a multifaceted phenomenon that identifies a set of those conditions and characteristics that enable an individual, or community to respond and recover from disasters.

Baba and Abayomi (2013) conducted a similar study in Ghana which assessed the effects of seasonal floods on livelihoods and food security. The study also analyzed the definition of the Inter-Governmental Panel on Climate Change-IPCC (2014) that vulnerability is the susceptibility or propensity of a system to be negatively affected by an external hazard. It discussed the causes of floods and damages and their impacts on food security and livelihood. The susceptibility exposure of farmland, agriculture crops and soil nutrients was significantly affected which contributed to low crop yield and food security.

Vink and Takeuchi (2013) conducted a study to evaluate legal actions in Disaster Risk Management (DRM) taken for vulnerable people in Japan, the Netherlands and the USA. The authors mentioned that in general some characteristics and sets of conditions make people more susceptible to disasters in a community despite actions taken on account of recovery or reducing disaster effects. However, in the study countries' definitions of vulnerable people/groups were not similar to each other as per respective DRM laws and policies. In the respective laws of the USA and Japan, some characteristics of vulnerable groups were found similar which were elderly people, children and people with disability. Whereas, the Netherlands has mentioned that not self-reliant people were vulnerable to the laws of DRM. The varying definition of vulnerable people sets different trends to approach vulnerable people in respective countries. Thus, it was difficult to set standards and trends for measuring vulnerability in DRM laws. This study referred to Hyogo Framework for Action-HFA (2005) that stressed the development of indicators and indices which would evaluate the vulnerability of different groups.

Highfield et al. (2014) in the study of mitigation planning for hazard exposure, structural and social vulnerability, underscored that communities living near rivers were more exposed to flood hazards. Therefore, the authors stressed the need to prioritize high impact zones and take mitigation and recovery measures to avoid losses and damages in such areas.

Wisner et al. (2014) have explained the relationship between vulnerability and disaster through the "pressure and release model". The model focus on the socioeconomic pressure and physical exposure that generate disasters. The division of upper class, lower class and marginalized class/groups in a community determine unequal access to resources which create unsafe conditions and dynamic pressures and further make them susceptible to the negative impact of natural events.

Tapia et al. (2017) conducted an indicator-based study to assess the vulnerability of 571 European cities to climate change. There were three risk factors for assessing climate change vulnerability which included flood, heat waves and drought. The information taken from the European audit database helped the development of indicators. The result presented that the vulnerability of the large cities was high due to the three risk factors of flood, heat wave and drought.

Busby asserted that mapping remains one of the significant approaches to help understand the degrees of prevailing vulnerability while conducting vulnerability assessments (Busby et al., 2013). The study analyzed degrees of spatial vulnerability in the flood hazard maps of target union councils. Sullivan-Wiley et al. (2019) used a participatory approach for developing vulnerability. The feedback of different stakeholders in the participatory planning contributed to understanding the community perception and validating vulnerable locations in the flood hazard map.

Dintwa et al. (2019) carried out a study to quantify the prevailing vulnerability in Botswana. Many indicators were used for measuring the vulnerability of people and places. The vulnerability index scores found that the highly vulnerable Districts (Ngamiland West and Central Tutume) were due to low income, unemployment, social insecurity, low education level and a high number of children and elderly people.

4.2 Studies Concerning Pakistan

The Asian Development Bank-ADB (2010) conducted post-flood need assessment of the flood affected areas in Pakistan. The report presented the extent of damages and losses in different sectors throughout Pakistan. It also provided insight into the gaps found in the preparedness and response measures. The report declared that the flood 2010 was the most devastating in the history of Pakistan (Bank, 2010). However, the responsible institution couldn't timely disseminate flood information, especially in the upper zone of Pakistan (Qasim et al., 2016). Government organizations also failed to evacuate people to a safer zone. The report recommended PKR 758 billion (US$ 8.9 billion) for rehabilitation and reconstruction needs.

Hong et al. (2011) conducted a study on Pakistan flooding 2010 by examining the European, and Russian blocking and the interaction of tropical monsoon surges. The study presented that a number of monsoon surges caused intense rainfall in Pakistan and triggered floods that devastated the country by affecting 20 million people with a death toll nearing 3000 individuals.

Mahmood and Ullah (2016) conducted a study to assess the flood 2010 causes and damages in Dir Valley Khyber Pakhtunkhwa. This study discussed the reasons for intense rainfall leading to floods. The floods' interaction with people's vulnerability in the area resulted in massive damages and losses to the socioeconomic and built environment. Zone-wise (Balkot, Barikot and Sharingal) damages to agriculture, infrastructure and economic losses were presented in this study. The assessment of flood damages emphasized the vulnerability of District Swat which is adjacent to Dir and has similar geographic features.

To examine resilience to flood, Qasim et al. (2016) conducted a community resilience assessment in District Charsadda, Nowshera and Peshawar. The result presented that all three sites were less resilient due to physical, social, economic and institutional factors which influenced community vulnerability. The study recommended policymakers take concrete measures for increasing the resilience of flood-prone people in the selected Districts.

Sadia et al. (2016) examined women's health risks and vulnerability in relation to floods in Pakistan. The floods 2010 most affected two UCs in District Nowshera were purposively selected. The findings presented that women were found more vulnerable to access better healthcare assistance. Multiple factors including low education, socially fabricated roles, access to extended services, income and perception of disaster make them more vulnerable in terms of public health. Flanagan et al. (2011)

also emphasized the provision of early health services in post-disaster situations. The limited available health services often led to increased mortality and prolonged recovery from disasters.

Climate change has widespread impacts on human beings with impacts varying from individual to individual. Abbasi et al. (2019) assessed gender vulnerabilities to climate change in three sub-basins of the Indus including Soan, Hunza and Chaj Doab in Pakistan. The findings revealed that climate change has left drastic impacts on the entire community of the study area. However, women and children were differently affected by extreme weather events. Women were found more vulnerable due to limited access to power, education, information, extension services and different socially fabricated roles and responsibilities. The rain-fed agriculture was the only source of income generation and food security. Climate change left negative impacts on rain-fed agriculture. Therefore, the men were compelled to search for livelihood sources in other areas and consequently, women had to take additional responsibilities in their absence. The study recommended that the government and aid agencies should provide financial assistance to the farmer community so that they could invest in climate resilient crops and diversified livelihood sources.

Rehman et al. (2019) conducted a study on the disaster management system in Pakistan. An example from the flood 2010 was narrated as a piece of evidence to justify the fragility of the system which was unable to address the root causes of vulnerability. The lack of coordination and working in isolation worsened the situation in flood 2010. The responsible institutions had no plan in place for an effective response.

Jamshed et al. (2019) conducted an empirical study to assess the relationship between capacity and vulnerability regarding flood in two Districts—Muzaffargarh and Jhang Punjab Pakistan. For assessing the vulnerability and capacity in realistic terms, a vulnerability index was developed. It was revealed that the findings of the vulnerability assessment were the same for the two Districts. However, different findings of capacity assessment portrayed that Muzaffargarh had more coping capacity to flood in terms of good education, access to resources, livelihood skills, social assets and community networking as compared to Sub-District Jhang.

5 Vulnerability Assessment of Study Area

The total population of Tehsil Behrain is 248,474 individuals (census 2017–2018). The target groups include men, women, people with disabilities, children, elected representatives, the small business community and government representatives. The study was conducted in four UCs—Madyan, Beshigram, Behrain and Tirat of Tehsil Behrain District Swat Khyber Pakhtunkhwa. The catchment area of District Swat is diverse for the reason that glaciers are melting in the northern area of the District, all making their way through the upper catchment. Similarly, monsoons also cause heavy rainfall. Specifically, from 27 to 28th July 2010, heavy rainfall led to a major flood after 80 years in Pakistan.

6 Data Source and Development of Vulnerability Index

The available literature was reviewed for the development of a vulnerability index which was developed comprised of three parameters exposure, fragility and lack of resilience. Indicators were selected for each parameter. In total, nineteen relevant indicators were selected. Three indicators were selected for exposure, eight each for the fragility and lack of resilience (Jamshed et al., 2019). Each indicator was further weighted with three class/questions on a scale of a high, medium, low and numerically as 3, 2 and 1, respectively. Vulnerability Index-VI was developed from the previous study including Hahn et al. (2009), Cutter et al. (2012), and Jamshed et al. (2019). Table 1, is presenting the vulnerability index with the aforementioned parameters and indicators rationalized in the given scale.

In total, 19 indicators were developed with 57 sub-indicators used for assessing people's exposures to hazards, fragility and resilience on a comprehensive scale of high, medium and low. Community exposure to hazards was identified through indicators of three priority hazards including flood, heavy rainfall and flash flood. Similarly, the fragile condition of the community and lack of resilience were evaluated by assessing the indicators of access to clean water, food security, health assistance, house pattern, community support structure, education, family income and environment/natural resources.

7 Measuring Flood Vulnerability by the Parameters Exposure, Fragility and Lack of Resilience

According to Smit and Wandel (2006), vulnerability is a function of three components: exposure to hazards, sensitivity to hazards and ability (capacity) to cope with hazards. The study assessed three parameters: community exposure to hazards, community's fragility and community's lack of resilience. The fragility of a system is all those unsafe conditions that make a system susceptible to flood. Similarly, lack of resilience reflects the underlying causes that reduce the system's capacity to withstand flood disasters.

The vulnerability of respective UCs was assessed. The information is entered into the database. The following equations/criteria are used to carry out an analysis of exposure, fragility and lack of resilience and identify Union Council Vulnerability.

$$\text{UC Vulnerability} = \text{Exposure} + \text{Fragility} + \text{Lack of Resilience}$$

where exposure, fragility and lack of resilience are calculated using the following equations:

$$\text{Exposure} = \sum_{i}^{n} \begin{array}{l}\text{Add exposure indicators Score} - \text{No of total} \\ \text{indicators used in exposure}\end{array}$$

Table 1 Parameters, indicators weighted questions on scale 3,2,1 as high, medium and low, respectively

Parameters	Indicators	Weighted Class/Questions	Scale
Exposure of community to hazards (A)	1- Priority Hazard (Flood)	More than two-thirds of the sub-communities exposed	3
		Between one-third and two-thirds of the sub-communities exposed	2
		Less than one-third of the sub-communities exposed	1
	2- Priority Hazard (Heavy Rainfall)	More than two-thirds of the sub-communities exposed	3
		Between one-third and two-thirds of the sub-communities exposed	2
		Less than one-third of the sub-communities exposed	1
	3- Priority Hazard (Flash Flood)	More than two-thirds of the sub-communities exposed	3
		Between one-third and two-thirds of the sub-communities exposed	2
		Less than one-third of the sub-communities exposed	1
Subtotal (A)		**Total for section A divided by the number of indicators in section A**	
Fragility (B)	1- Access to Clean Water	Clean water is not available to most (needs treatment)	3
		Clean water is available but not for all	2
		Clean water is always available locally	1
	2- Food Security	Day-to-day survival practice in the community	3
		Some reserves of essential items kept (e.g grain)	2
		Reserve stocks covering all basic needs are available (e.g grain + seeds + food stocks until the next season)	1
	3 -Health Assistance	No Lady Health Worker (LHW) and no locally accessible health assistance (within 1 days travel)	3
		Limited health assistance for the community (i.e. LHW)	2
		Basic accessible health facilities available locally (i.e. BHU/Clinic) with 1 days travel)	1
	4 - Community Support Structures	No system provides the required support	3
		A system able to provide partial support	2
		A strong system able to provide the required support	1
	5- Housing	House poorly constructed and made only from traditional local materials – kacha	3
		Basic Block House (brick and mud mortar) – pakka	2
		Brick & mortar or concrete housing with no building codes or structural reinforcement	1
	6- Education	No education facility within a 1 Km distance	3
		Primary Education facility available within 1 Km distance	2
		Primary and Secondary Education facilities available locally	1
	7- Basic Income	Regular means of income from one source of income (e.g. Agriculture for a limited number of crops)	3
		Regular incomes from one source and irregular incomes from different sources (inside and outside of the community)	2
		Regular means of income from different sources (including outside of the community)	1

(continued)

Table 1 (continued)

	8- Environment / Natural Resources	No/very limited availability of some essential natural resources	3
		Limited local availability of required natural resources (pasture land, irrigation water, cultivation land limited accessibility)	2
		Local availability of required natural resources (pasture land, rain-fed cultivation land, forest accessible for all)	1
Subtotal (B)		**Total for section B divided by the number of indicators in section B**	
Lack of Resilience (C)	**1- Access to clean water**	Lacking relevant skills and resources to make repairs	3
		Some of the required skills & materials available to make repairs (Basic WASH O&M)	2
		Required skills & materials available to make repairs (WASH O&M)	1
	2- Food Security / Assets	Assets uncommon in the community	3
		Assets available in the community (e.g. livestock)	2
		Saving schemes utilised (banks, community saving schemes, etc)	1
	3 - Health Assistance	The community lacks skills and knowledge for treating common diseases / health problems	3
		Effective community skills and knowledge for treating common diseases / health problems	2
		Effective community skills and knowledge for treating most diseases / health problems	1
	4 - Community Support Structures/ Preparedness	No community support system (dependent on outside systems)	3
		Part of the support system run by the community	2
		The community has its mechanisms for responding and providing support	1
	5- Housing	Lacking relevant skills and resources to make repairs	3
		Some of the required skills & materials available to make repairs	2
		Required skills & materials available to make repairs	1
	6- Education	Available Education facilities are unsafe	3
		Available Education facilities in safe locations	2
		Available Education facilities in safe locations with safe construction	1
	8 –Basic Income	No/very low education and vocational skills within the community	3
		Low education and vocational skills within the community	2
		Education and vocational skills common within the community	1
	8 - Environment/ Natural Resources	Natural resources not managed	3
		Some protection / management of some natural resources	2
		All essential natural resources protected (e.g. forest protection committee, NRM Committee, etc)	1
Subtotal (C)		**Total for section C divided by the number of indicators in section C**	
VULNERABILITY		**A + B + C**	

Table 2 Vulnerability scoring criteria

Categories	Community vulnerability category	Criteria, vulnerability score range
Low	1	1–1.49
	2	1.5–2.49
	3	2.5–3.49
Medium	4	3.5–4.49
	5	4.5–5.49
	6	5.5–6.49
High	7	6.5–7.49
	8	7.5–8.49
	9	8.5–9

$$\text{Fragility} = \sum_{i}^{n} \text{Add fragility indicators Score} - \text{No of total indicators used in fragility}$$

$$\text{Lack of Resilience} = \sum_{i}^{n} \text{Add Lack of resilience indicators Score} - \text{No of total indicator used in LR}$$

The maximum vulnerability score was calculated as 9 in the vulnerability index, whereas, the lowest minimum vulnerability was calculated as 1. Table 2 explains Vulnerability Scoring Criteria.

8 Assessment and Analysis of Community Exposure, Fragility and Lack of Resilience

The community prioritized that three hazards namely, major flood, heavy rainfall and flash floods had adversely impacted their socioeconomic infrastructure. Thus, the study considers only the top 3 hazards present in the community. The response of the community's exposure was recorded in the ranks from 1 to 3 (1 = LOW, 2 = MEDIUM, 3 = HIGH) and the score was identified for three priority hazards of underexposure. The score of three priority hazards was summed and divided by the total number of indicators. For instance: a respondent response was identified as 02 for major flood, 1 for flash flood and 3 for heavy rainfall. All three indicators were summed as $2 + 3 + 1 = 6$. Then, 6 was divided by 3 (total number of indicators) to get the total exposure score of a respondent. Subsequently, the community Exposure Score calculated was 2. Thus, we used the following equations to identify mean exposure to hazards (flood, heavy rainfall and flash flood).

$$\text{Exposure} = \sum_{i}^{n} \text{Add exposure indicators Score} - \text{No of total indicators used in exposure}$$

Similar to exposure, here too, the indicators developed for the fragility and lack of resilience were ranked as high-3, medium-2 and low-1. The abovementioned computation process was repeated for each response to get a score of fragility and lack of resilience, respectively. For example: Clean Water Access = 2, Housing = 3, Education = 2, Basic Income = 2, Food Security = 1, Health = 2, Total: 2 + 3 + 2 + 2 + 1 + 2 = 12. Divide 12 by 6 (number of indicators) = 2. In this way, Community Fragility Score was identified. The following equation is used for identifying community fragility.

$$\text{Fragility} = \sum_{i}^{n} \text{Add fragility indicators Score} - \text{No of total indicators used in fragility}$$

For lack of resilience, we supposed that:

Clean Water Access = 3, Housing = 3, Education = 2, Basic Income = 2, Food Security = 2, Health = 3, Total: 3 + 3 + 2 + 2 + 2 + 3 + 2 + 3 = 20. Divide 15 by 8 (number of indicators) = 2.5. Thus, the given equation is used for figuring out the lack of resilience.

$$\text{Lack of Resilience} = \sum_{i}^{n} \text{Add Lack of resilience indicators Score} - \text{No of total indicator used in LR}$$

9 Analysis of Community Vulnerability

The study used a vulnerability index to assess vulnerability by analyzing the primary and secondary information. The analysis identified that the indicators used for exposure, fragility and lack of resilience show consistency in the findings of respective union council vulnerabilities. Several studies including District Swat Disaster Management Plan and Tehsil Behrain Disaster Management Plan developed by Provincial Disaster Management Authority gave only a brief overview of vulnerability in the District or study area. In District Disaster Management Plan, Tehsil Behrain has been categorized as low-vulnerable Tehsil whereas the analysis of our study revealed all four union councils are highly vulnerable and therefore, an extension of research to the remaining union councils is recommended. This reveals a policy gap for the decision-makers and suggests a serious need to revisit their plan's findings and change the status of Tehsil Behrain as highly vulnerable Tehsil is prone to multiple disaster effects.

The analysis of exposure component/variable/parameter gives the insight that a major portion of the community is exposed to the effects of flood hazards and the Union Council Madyan is highly exposed to the priority hazards including major floods, flash floods and heavy rainfall. Capacity associated with exposure is limited as the area lacks retaining walls, check dams and vegetation cover which can deter disaster effects in times of flood disasters.

The findings revealed that the majority of the population relies on spring water which is neither protected nor free from contamination. In rural areas, women/girls ply long distances to pitch water from the springs mostly located in faraway mountains. The analysis further reveals that such activities significantly affect girls' education. In urban areas, clean drinking water was channelized to houses by PVC pipes which often pass through drains leading to contamination of drinking water.

Food security is another major factor in people's increasing vulnerability. The local people are mainly dependent on persistent agriculture yields that hardly meet survival needs or merely cover basic needs. The results reflect that people lack the capacity to use advanced technology for increasing their yields and production from agriculture and livestock.

The fragility in health assistance identifies that limited people have access to basic health facilities. In case of serious health issues, people visit the hospital in down Districts. In the time of disasters, limited medical facilities and staff are not able to provide health assistance to people affected by disasters.

The findings showed that recurrent disasters have weakened local support structure and networking. The isolation in thinking and actions significantly disturb the social cohesion in the community. The findings indicate that fund generation and collective actions for resolving communal problems do not exist at any level in the community.

The housing structures are vulnerable and are unable to withstand flood disasters. People lack capacity in terms of building resilient structures by following building regulations, technology and techniques. It has been observed that people are reconstructing buildings in flood-prone areas which indicate low awareness about disaster risk, and no implementation of land regulations by government departments.

The available education facilities were unsafe and beyond access for girls especially. Therefore, girls' education was very low as indicated by respective results. Moreover, the education facilities were located in flood-prone areas and not resilient to withstand floods, earthquakes and rock falls (from nearby mountains).

Income from tourism was the 2nd largest source after income from the agriculture sector. At one time, women were participating equally along with their men to generate income from embroidery and handmade stuff. They had good linkages with local markets and even down District markets. The recurrent disasters including the flood 2010 disrupted the market setup and tourism.

The findings revealed the vegetation cover and watershed were also badly damaged by the flood 2010. However, it has been noted that forest covers were drastically affected by the timber business, household use of wood as fuel and abundant use in building infrastructure.

10 Recommendations

The study has laid down baseline information for devising mitigation plans and conducting further research studies on vulnerability to flooding.

- Construction of dikes, check dams and vegetation cover is recommended for reducing vulnerability in the study area.
- Tehsil Behrain has been declared a low-vulnerable Tehsil in the Disaster Management Plan of Disaster Swat by PDMA. Therefore, its status needs to be a highly vulnerable Tehsil prone to multiple disaster effects.
- Women being a sensitive figure in the community needs special attention in terms of health care, financial support, protection and psychosocial support.
- Programs should be designed for the strengthening of the community support structure and social cohesion by engaging the local community and making their linkages with government and nongovernment stakeholders.
- Implementation of land regulation is vital to avoid construction in flood-prone areas.
- The directorate of education is recommended to take concrete actions for making education facilities accessible and disaster resilient. Special education programs should be designed for those girls who left education due to recurrent disasters.
- Revitalization of women entrepreneurship with adequate access to the market within cultural boundaries.
- Local people need to be sensitized about the growing effects of climate change and should be prepared for meeting these effects through adaptation measures. Similarly, farmer's training on diversified livelihood sources needs to be imparted particularly women capacity building may be ensured.

11 Conclusions

The study evaluated the community vulnerability of four UCs of Swat districts against hazards prioritized by the community. Major floods, heavy rainfall and flash floods were identified as top-priority hazards. The hazards were assessed against exposure, fragility and lack of resilience for quantifying community vulnerability. Findings revealed that vulnerable groups including women were highly affected due to recurrent disasters, especially girls' education was disrupted and needs attention. Due to a lack of awareness and land use regulation implementation issues, people again started building infrastructures showing complete negligence of build back better options. The second major source of income which was tourism was severely affected in the aftermath of disasters. To address these issues, the government and concerned stakeholders need to ensure immediate disaster risk reduction measures at the community level for capacity building and strengthening institutional setup through public-private partnerships.

References

Abbasi, S. S., Anwar, M. Z., Habib, N., Khan, Q., & Waqar, K. (2019). Identifying gender vulnerabilities in context of climate change in Indus basin. *Environmental Development, 31*, 34–42.

Baba, A., & Abayomi, O. (2013). Effects of seasonal floods on households' livelihoods and food security in Tolon/Kumbumgu district of the Northern Region, Ghana. *American Journal of Research Communication, 1*, 160–171.

Bank, W. (2010). *Pakistan floods 2010: Preliminary damage and needs assessment.* World Bank.

Busby, J. W., Smith, T. G., White, K. L., & Strange, S. M. (2013). Climate change and insecurity: Mapping vulnerability in Africa. *International Security, 37*(4), 132–172.

Cutter, S. L., Boruff, B. J., & Shirley, W. L. (2012). Social vulnerability to environmental hazards. In S. L. Cutter (Ed.), *Hazards vulnerability and environmental justice* (pp. 143–160). Routledge.

Dintwa, K. F., Letamo, G., & Navaneetham, K. (2019). Quantifying social vulnerability to natural hazards in Botswana: An application of cutter model. *International Journal of Disaster Risk Reduction, 37*, 101189.

Fekete, A. (2009). Validation of a social vulnerability index in context to river-floods in Germany. *Natural Hazards and Earth System Sciences, 9*(2), 393–403.

Flanagan, B. E., Gregory, E. W., Hallisey, E. J., Heitgerd, J. L., & Lewis, B. (2011). A social vulnerability index for disaster management. *Journal of Homeland Security and Emergency Management, 8*(1), 1–22.

Hahn, M. B., Riederer, A. M., & Foster, S. O. (2009). The livelihood vulnerability index: A pragmatic approach to assessing risks from climate variability and change—A case study in Mozambique. *Global Environmental Change, 19*(1), 74–88.

Highfield, W. E., Peacock, W. G., & Van Zandt, S. (2014). Mitigation planning: Why hazard exposure, structural vulnerability, and social vulnerability matter. *Journal of Planning Education and Research, 34*(3), 287–300.

Hong, C. C., Hsu, H. H., Lin, N. H., & Chiu, H. (2011). Roles of European blocking and tropical-extratropical interaction in the 2010 Pakistan flooding. *Geophysical Research Letters, 38*(13). https://doi.org/10.1029/2011GL047583

Jamshed, A., Rana, I. A., Mirza, U. M., & Birkmann, J. (2019). Assessing relationship between vulnerability and capacity: An empirical study on rural flooding in Pakistan. *International Journal of Disaster Risk Reduction, 36*, 101109.

Mahmood, S., & Ullah, S. (2016). Assessment of 2010 flash flood causes and associated damages in Dir Valley, Khyber Pakhtunkhwa Pakistan. *International Journal of Disaster Risk Reduction, 16*, 215–223.

Morrow, B. H. (1999). Identifying and mapping community vulnerability. *Disasters, 23*(1), 1–18.

Mwape, Y. P. (2009). *An impact of floods on the socio-economic livelihoods of people: A case study of Sikaunzwe community in Kazungula District of Zambia* (Unpublished Dissertation for Award of MSc Degree at University of the Free State). 87pp.

Qasim, S., Qasim, M., Shrestha, R. P., Khan, A. N., Tun, K., & Ashraf, M. (2016). Community resilience to flood hazards in Khyber Pukhthunkhwa province of Pakistan. *International Journal of Disaster Risk Reduction, 18*, 100–106.

Rehman, J., Sohaib, O., Asif, M., & Pradhan, B. (2019). Applying systems thinking to flood disaster management for a sustainable development. *International Journal of Disaster Risk Reduction, 36*, 101101.

Sadia, H., Iqbal, M. J., Ahmad, J., Ali, A., & Ahmad, A. (2016). Gender-sensitive public health risks and vulnerabilities' assessment with reference to floods in Pakistan. *International Journal of Disaster Risk Reduction, 19*, 47–56.

Smit, B., & Wandel, J. (2006). Adaptation, adaptive capacity and vulnerability. *Global Environmental Change, 16*(3), 282–292.

Sullivan-Wiley, K. A., Gianotti, A. G. S., & Connors, J. P. C. (2019). Mapping vulnerability: Opportunities and limitations of participatory community mapping. *Applied Geography, 105*, 47–57.

Tapia, C., Abajo, B., Feliu, E., Mendizabal, M., Martinez, J. A., Fernández, J. G., Laburu, T., & Lejarazu, A. (2017). Profiling urban vulnerabilities to climate change: An indicator-based vulnerability assessment for European cities. *Ecological Indicators, 78*, 142–155. https://doi.org/10.1016/j.ecolind.2017.02.040

Vink, K., & Takeuchi, K. (2013). International comparison of measures taken for vulnerable people in disaster risk management laws. *International Journal of Disaster Risk Reduction, 4*, 63–70.

Wisner, B., Blaikie, P., Cannon, T., & Davis, I. (2014). *At risk: Natural hazards, people's vulnerability and disasters*. Routledge.

Socio-Economic Potentials of Nigerian Inland Waterways System for Sustainable Use

Solomon Usman Jatau, Araoye Olarinkoye Ajiboye, and Babatope Gabriel Oni

1 Introduction

Transportation is vital to any economy as it is inseparably linked to other sectors of the economy. It is the vehicle for inter- and intra-country trade and the means of movement include air, pipeline, rail, ocean or Inland Waterways. Inland Waterways are any navigable body of water either natural or manmade such as rivers, canals or routes used for travel by water from the coast or shoreline to the interiors of a country. Inland water transportation (IWT) is, therefore, the utilisation of Inland Waterways for the movement of passengers and freight between inland ports, wharves or quays. Like any other mode or class of transportation, it facilitates the social and economic interaction of citizens and serves as a major catalyst for national development.

According to Lawal and Oluwatoyin (2011), countries with vast Inland Waterways have utilised it as an important factor in the development of their regions through oil exploration, the opening of new settlements and commerce and trade. Water transport being one of the cheapest modes of transport has the largest carrying capacity and is most suitable for carrying bulky goods over long distances. The development of navigable Inland Waterways has equally assisted nations to achieve several developmental goals and offers the most economical, energy efficient and environmentally friendly means of transporting passengers and freight. In most cases, it requires much less capital investment and maintenance than other transport modes. Despite evident lesser environmental impacts when compared to other competing modes, they have been rather neglected in many countries including Nigeria.

S. U. Jatau (✉) · A. O. Ajiboye · B. G. Oni
Department of Logistics and Transport Technology, School of Innovative Technology, Federal University of Technology Minna, Minna, Nigeria
e-mail: solojatau2002@yahoo.com

A. O. Ajiboye
e-mail: araoyeoajiboye@futminna.edu.ng

N. Shahzad (ed.), *Water and Environment for Sustainability*,
https://doi.org/10.1007/978-3-031-27280-6_13

Nigeria has the second longest length of Inland Waterways in Africa after Egypt. It has 8,600 km of Inland Waterways centred on River Niger and River Benue and their tributaries (Ndikom, 2008). The coastal waterways extend from Lagos State through Warri in Delta State to Calabar in Cross River State. The coastal ways from Lagos through the River Niger link up the Eastern part of the Country. The IWT operation is advantageous in terms of the costs of moving heavy traffic by road transport with an average share of about 1.6% of Nigeria's GDP. A single 15-barge load, for instance, is comparable to around 225 rail cars or 870 tractor-trailers. The transportation of tonnes of agricultural produce and imported items would gain most from this.

The Federal Government of Nigeria (FGN) established the National Inland Waterways Authority (NIWA) in 1997 as part of measures to promote Inland Waterway transport in Nigeria. According to Usman and Oyesiku (2015), the main responsibility of NIWA is to improve and develop Inland Waterways for navigation and provide an alternative mode of transportation of economic goods and persons among other things. NIWA has been able to partially dredge the lower Niger at Onitsha and has built inland ports at Lokoja, Baro, Oguta and Onitsha while the contract for the dredging of River Benue is under consideration. Most recently, the FGN is currently making efforts to dredge the Inland Waterway leading to the Warri Port as part of efforts to decongest the Apapa port which is the major challenge to the Lagos State traffic gridlock. Despite these achievements, IWT in Nigeria is underutilised and underdeveloped. As a result of the poor utilisation of water resources in Nigeria especially that of River Niger, the Federal Government of Nigeria (FGN) looses a lot of revenue which could accrue to boost the economic activities of the country.

2 An Analysis of Studies on Inland Water Transportation in Nigeria

The late Professor R. K. Udo carried out groundbreaking research on inland water transport in Nigeria in the 1970s. Water is one of the many natural resources that Nigeria possesses, according to Udo (1970), and the nation has the opportunity to provide the majority of West Africa's landlocked nations, including Burkina Faso, Chad, Mali and Niger, for greater economic gains. Iloje (1984) also noted that approximately 8000 kms of Nigeria's internal rivers are navigable and that the country is highly endowed with surface water resources.

Many other researchers, including Badejo (1997), have published on numerous IWT-related topics in Nigeria, including their history, benefits, management challenges and potentials. For instance, Badejo (1997) and Adams (1998) determined that the country's namesake river, the Niger, and its major tributary, the Benue, are the principal rivers whose channels provide the longest waterways into the country's hinterland. The fact that both rivers originate outside of the nation yet meet at the confluence of Lokoja and later flow into the Gulf of Guinea via a vast network of distributaries and creeks that make up the Niger Delta emphasises the significance of these rivers for IWT.

Adams (1999) found that Nigeria now has navigable canals with a capacity of around 10,000 kms and a long coastline of roughly 852 kms. On the basis of this, he observed that the nation had a considerable potential for moving cargo and people by water from the coast to the inland. Anyam (2003) lamented the fact that potential investors have not yet taken advantage of the enormous business opportunities that IWT offers, which may enable it to compete with both road and air travel. Adams (2004) supported this by lamenting that the FGN had not yet given IWT the attention it deserved, notably in terms of finance and the building of infrastructure. The lack of channelisation and dredging of navigable rivers, the poor construction and rehabilitation of river ports, the scarcity of infrastructure for water transportation and safety and security issues along the navigable waterways were some of the barriers identified to a sustainable IWT.

In contrast to Bangladesh, where water transport accounts for 32% of the transport sector (Rahman, 1994), the Philippines, where it accounts for 20% of the sector (Fellinda, 2006), Sierra Leone, where it accounts for 3% of the sector (Kimba, 2008), India, where it accounts for 0.15% of the sector (Raphuram, 2004) and Nigeria, where it accounts for only 0.08%, Ezenwaji (2010) focused on the underutilisation of Inland Waterways in Nigeria (NIWA). Like earlier scholars Aderamo and Mogaji (2010), he found that a number of natural elements have a negative impact on Nigeria's use of inland rivers as transportation routes.

Ndikom (2013) found that IWT gave investors opportunities such as facility management, jetty operations and boat building. He noted that issues of security concerns discouraged potential and foreign investors. River channel dredging and maintenance, a lack of private sector participation in the water transport industry, the construction and rehabilitation of river ports, the purchase of passenger ferries, security boats, the construction of channel buoys and other projects were other problems limiting IWT's viability. Conflicts between federal and state agencies overseeing IWT in Nigeria were more concerning, and they have stymied the growth of that industry.

The above review underpins that the state of IWT in Nigeria is still largely underdeveloped despite efforts by stakeholders. The operation, its major contributions, socio-cultural constraints limiting its operations and policy trusts and targets, etc. remain largely unknown. However, none of them examined the challenges militating against the utilisation of the River Niger nor proffered strategies to mitigate the challenges. It is this gap in the literature that this study seeks to fill by appraising the utilisation of the River Niger Inland Waterways for enhanced economic and national development in Nigeria.

3 Issues on Nigerian Inland Waterways System

The Inland Waterways have been a major source of transport conveying both passenger and freight even before the independence of Nigeria. In particular, IWT has accounted for well over 38% of produce transported in the country. Commodity transportation across the riverine and coastal state of the federation has been increasing from the early 1960s to date. By implication, IWT has made a significant contribution to the economic and national development of the country.

However, the demand for transport is growing in Nigeria, which may require the development of the nation's Inland Waterways for navigation. In particular, environmental changes and pressure from other human activities have resulted in little priority accorded to Inland Waterways and thus reduce the viability of the river for IWT. In addition, previous administration has seen the need to improve the viability of the Inland Waterways system, especially the Rivers Niger and Benue. This is due to her prospect of promoting commodity flow along the riverine areas and national development.

According to Table 1, across the years studied, the highest quantity of 7.91 tonnes of agricultural produce was transported through Inland Waterways in 2021, closely followed by 7.48 tonnes in 2020 and in 2019, 7.39 tonnes was transported. Similarly, 32.8 tonnes of cassava was the highest produce transported across 10 years, followed by yam with 25.2 tonnes moved and 8.1 tonnes of palm kernels.

Given Nigeria's highly limited and congested transport environment, one major shortcoming of IWT development in Nigeria is the non-sustainability of IWT services by different levels of governments and the lack of private sector participation in Nigeria which has resulted in the ugly state of IWT irrespective of its glaring importance in the development of the economy. For instance, traders in Aba and Onitsha, which both have robust industries producing shoes, apparel, textiles, plastics, pharmaceuticals, petrochemicals, polyethylene, cosmetics and aluminium products, are big importers and exporters. On a daily basis, the Onitsha market receives more than five million visitors and conducts over twelve million unique transactions. Over $3 billion in trade is transacted annually at the Onitsha market, with unbanked transactions accounting for about 40% of this total. This translates to Onitsha being one of Nigeria's most economically significant cities and having one of the highest GDPs. Recently the Onitsha inland port is now operational, especially with the dredging of River Niger which makes it navigable in a bid to covey commodities to Onitsha thus decongesting the Lagos seaport for economic and national development of the country.

Table 2 further shows the high demand for the movement of containerised cargoes across the Nigerian Inland Waterways system, with the Lokoja area leading, followed by Onitsha and Makurdi areas. This is very important for informing business decisions by investors. Also, it shows that no significant relationship exists between Warri, Onitsha and Makurdi in terms of TEUs handled, while Lokoja was found 100% different from the three stations. Also, no significant relationship exists between

Table 1 Statistics of generated agricultural produce transported through Inland Waterways 2012–2021

Products	2012 (tonnes)	2013 (tonnes)	2014 (tonnes)	2015 (tonnes)	2016 (tonnes)	2017 (tonnes)	2018 (tonnes)	2019 (tonnes)	2020 (tonnes)	2021 (tonnes)	Mean	Ranking
Beans	1.8	1.8	1.9	2.1	2.1	2.2	2.2	2.2	2.2	2.3	**2.1**	**11**
Cassava	31.4	31.4	32.1	32.7	32.7	32	32.1	32.7	32.9	38.2	**32.8**	**1**
Groundnuts	1.6	2.3	2.5	2.5	2.9	2.9	2.7	2.7	2.8	2.9	**2.6**	**10**
Maize	6.9	5.7	5.3	5.1	5.5	4.1	4.6	4.7	4.8	4.8	**5.2**	**7**
Millet	5.6	5.7	5.9	6	6	6.1	5.5	6.1	6.3	6.3	**6.0**	**6**
Palm kernel	7.3	7.3	7.8	7.8	8	8.2	8.5	8.5	8.6	8.7	**8.1**	**5**
Plantain	1.6	1.7	1.7	1.8	1.9	2	2	2.1	2.1	2.1	**1.9**	**13**
Rice	2.9	3.1	3.3	3.3	3.3	3.3	2.8	3.2	3.4	3.5	**3.2**	**8**
Sorghum	7	7.1	7.3	7.5	7.5	7.7	7.1	7.7	8	8	**7.5**	**4**
Starchy roots	1.2	1.2	1.8	3.8	3.8	3.9	3.9	3.9	4	4	**3.2**	**8**
Sugar cane	0.6	0.6	0.7	0.7	0.7	0.7	0.7	0.7	0.7	0.8	**0.7**	**14**
Sweet potato	1.2	1.5	1.5	1.6	2.5	2.5	2.5	2.5	2.5	2.5	**2.1**	**11**
Wheat	–	–	0.1	0.1	0.1	0.1	0.1	0.1	0.1	0.1	**0.1**	**15**
Yam	22.8	23.2	24	24.8	25.9	26.2	26.2	26.3	26.3	26.6	**25.2**	**2**
Mean	**7.07**	**7.12**	**6.85**	**7.13**	**7.35**	**7.28**	**7.21**	**7.39**	**7.48**	**7.91**		
Ranking	**9**	**8**	**10**	**7**	**4**	**5**	**6**	**3**	**2**	**1**		

Source NIWA (2022)

Table 2 Data on the Potential of moving containerised cargo from Port Harcourt and Onne Ports by Inland Waterways, 2012–2021

Years	Warri			Onitsha			Lokoja			Makurdi		
	TEUs	Tons	% TEUs	TEUs	Tons	% TEUs	TEUs	Tons	% TEUs	TEUs	Tons	% TEUs
2012	1,600	36,000	**1.3**	9,800	102,000	**2.7**	36,500	456,000	**2.1**	2,400	40,500	**1.8**
2013	2,800	40,600	**2.2**	15,500	221,500	**4.2**	55,100	787,300	**3.1**	3,200	46,100	**2.4**
2014	3,330	48,000	**2.6**	24,000	302,500	**6.5**	77,600	966,000	**4.4**	44,500	79,000	**33.5**
2015	5,400	74,200	**4.3**	28,300	390,400	**7.7**	123,900	1,708,700	**7.0**	6,800	93,500	**5.1**
2016	60,500	88,000	**48.1**	31,000	464,500	**8.4**	166,000	2,001,500	**9.4**	7,200	122,000	**5.4**
2017	7,500	100,500	**6.0**	38,800	520,100	**10.5**	183,800	2,467,100	**10.4**	9,800	131,600	**7.4**
2018	8,800	102,800	**7.0**	42,000	599,500	**11.4**	211,000	288,000	**11.9**	11,200	176,000	**8.4**
2019	10,100	132,900	**8.0**	51,600	678,800	**14.0**	264,300	3,477,200	**14.9**	13,700	180,100	**10.3**
2020	12,500	156,400	**9.9**	60,500	702,300	**16.4**	288,000	421,500	**16.2**	15,600	196,600	**11.7**
2021	13,200	171,800	**10.5**	67,000	868,500	**18.2**	368,000	4,772,800	**20.7**	18,500	240,000	**13.9**
Mean	**12,573**	**95,120**		**36,850**	**485,010**		**177,420**	**1,734,610**		**13,290**	**130,540**	
Ranking	**4th**			**2nd**			**1st**			**3rd**		

Source NIWA (2022)

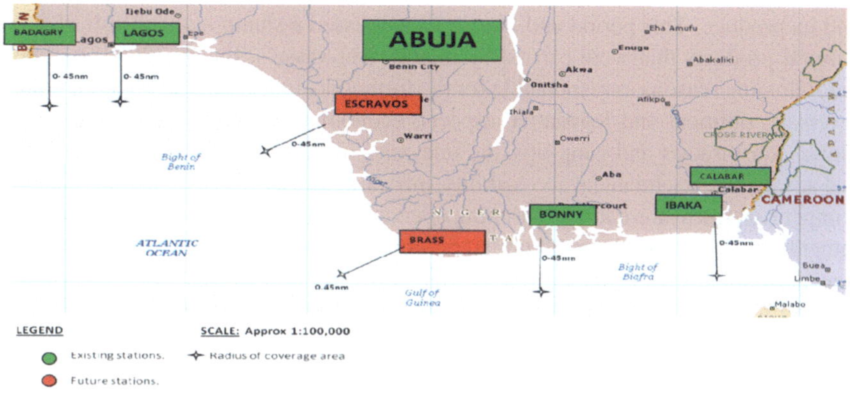

Fig. 1 Map of the study area

Warri, Onitsha and Makurdi in terms of capacity, while Lokoja was found 100% different from the three stations.

4 Research Scope

The scope of this study is delimited by time, space and content boundaries. In terms of time, the study covered the period from 2012 to 2021. In terms of space, the study focused on locations around the Niger and Benue rivers which include States like Kogi, Delta, Anambra and Abia. This is because these states utilise River Niger for IWT. In terms of content, the study is limited to the viability of the River Niger for IWT. This content was chosen because it links one of the most viable commercial cities along the Rivers Niger and Benue (Fig. 1).

5 IWT Development in Nigeria

There have been various attempts to build a practical IWT in Nigeria, as per the information obtained during the course of this study from pertinent literature and interviews. From the colonial to postcolonial era, the nation's marine infrastructure generally was developed in four phases, resulting in a total of 8 seaports with a projected combined transport capacity of 35 to 60 million tonnes of cargo yearly.

Some major events after Nigeria gained her independence in 1960 such as the breakout of the civil war in 1966 reduced the relevance of IWT and thus affected revenue and employment opportunities. The oil boom of the 1970s further reduced the use of IWT in Nigeria as dependence on oil revenue resulted in the massive development of road infrastructure with a drastic reduction in reliance on agriculture and its produce. This period and afterward witnessed a slump in the export of agricultural products that were carried by inland river vessels. Badejo (2009) outlined the following reasons as responsible for the further decline of IWT in Nigeria: policy reversals, political and legislative aspects, economic dynamism and market size, global conspiracies and gang-ups, the relevance of human capacity and technological advancements, financial accessibility, infrastructure and equipment failure, fierce competition from international shipping operators, a general lack of coordination and integration, multilateral versus unilateral competition and an overreliance on road transportation are all factors.

This trend continued throughout the 1980s and 1990s until the FGN set up NIWA through Decree No. 13 of 1997 in a bid to resuscitate IWT. The NIWA was granted the sole responsibility to manage, regulate and control all Inland Waterways in the country as well as ensure the development of infrastructural facilities for IWT as contained in the extract of NIWA Act 2004. Since its establishment, the IWT did not improve as most of the bottlenecks experienced before its establishment continued. Contemporary challenges such as insufficient funding, inadequate skilled force and worsening climate change coupled with human activities affected the nature of the rivers, as well as growing insecurity problems. Despite the challenges, NIWA achieved some strides through the construction of some facilities such as the inland river port at Akpanya near Lokoja, Kogi State and the construction of a jetty at Yenegoa II, Bayelsa State. A committee was set up in June 2019 and chaired by the Vice President, Prof Yemi Osinbajo, with a task to ease the gridlock experienced in Lagos as a result of trucks and trailers transporting containers from Apapa port to the other parts of the country. One of the decisions taken by the committee was to commence the construction of a port in Warri (Fig. 2).

Fig. 2 Bonny River (not navigable). *Source* Authors' field survey, 2021

6 Efforts Made by Federal Government in Sustaining IWT in Nigeria

1. **Establishment of NIWA**

After the country gained independence in 1960, IWT began to deteriorate. The Inland Waterways Department (IWD) kept a respectable amount of authority over the Inland Waterways network of the country, but took few management initiatives. The Federal Government of Nigeria established the NIWA through Decree No. 13 of 1997 to follow the disbanded IWD, with a clear mandate to manage Nigeria's extensive Inland Waterway resources. This was done in an effort to revive IWT in Nigeria and make it efficient, competitive and result-oriented. The NIWA Act of 2004 contained the vision statement: to make Nigeria the leader in the development and administration of inland water transportation in Africa, which is still far from being achieved. Its mission was to develop infrastructure facilities for an effective inter-modal transportation system in accordance with international best practises that is safe, seamless and reasonably priced. It also provided regulatory, economic and operational leadership in the country's Inland Waterways system.

NIWA was also saddled with the responsibility to provide the regulatory services range from giving permits and licences for sand dredging, pipeline construction, slot dredging, and approving designs and building of inland river crafts to inspecting and surveying inland watercraft and shipyard operators. Construction of inland river ports and jetties, capital and maintenance dredging, engineering design of river ports, hydrological and hydrographical surveys, river chart creation and cartography, river mapping, aerial survey and underwater survey were some of the additional services offered. In the inland water and its right of way, NIWA is also permitted to conduct an environmental impact assessment of dredging and other navigational activities.

2. **Dredging of the River Niger for IWT**

Since the days of colonial rule, plans have been made to dredge the lower Niger (Badejo, 2010). It has proven to be too difficult and frustrating to execute on numerous occasions, leading succeeding governments to cancel numerous contracts even after paying the contractors billions of Naira. The first proposal for the dredging of the River Niger was submitted by the then IWD and later the Petroleum Trust Fund (PTF) in 1996 but handed over the Lower Niger Dredging to the NIWA in November 2000. After 9 years of inactivity, the dredging of River Niger was flagged off on 10 September 2009 at a total cost of N38 billion. The project crystallised the then administration's determination to develop the nation's IWT sector. The 572-km-long dredging, according to Arc. Ahmed Yar'Adua, Managing Director of the Nigerian Inland Waterways Authority (NIWA), encompassed Baro in Niger State to Bifurcation in Bayelsa State as,

> the dredged river is expected to provide all-year round navigation, employment opportunities, and improved economic activities as well as flood control. Other benefits he says, include improved carrying capacity, cheaper and safer transport system, reduced axle load on roads, boom fishing and preserve the environment.

After the failure of the dredging of the Ikpoba River in Edo State and the Calabar port in Cross River State, he explained that it was the first significant dredging operation ever to take place in the nation. The dredged Lower River Niger covered 152 settlements in the states of Delta, Bayelsa, Rivers and Edo, as well as Imo, Anambra, Kogi and Niger.

Due to the nature of the River Niger which flows through the alluvial bed and like most alluvial river generates sediment that required continuous dredging. Coupled with the paucity of funds, and lack of commitment of successive administrations, the capital dredging could not be maintained. Dogara (2017) while describing how the Olusegun Obasanjo government created the national transport master plan, stated:

> I was part of the development of the national transportation master plan, I don't know what has happened to that, how I wish we implemented that plan, by now we would have gone very far because the national transportation master plan vision is a seamless integration of a multi-modal transportation into one hub, and I think at that time Baro in Niger State had been identified as the center for the hub, it is serviced by a railway that is connected to Minna and all you need to do is to construct a road and airport that will ensure connectivity and you will have access to the Atlantic Ocean through the River Niger, so everything was supposed to be put together in one place.

Table 3 shows that across all the stations studied, the highest water level of 7.3 m was recorded in the month of September, followed by 6.8 m in October and 6.0 m observed in August. The lowest water level of 2.5 m was observed in February and April, followed by 2.6 m recorded in January, March and May, respectively (Fig. 3).

All other pairs of waterways in Nigeria apart from the below combinations have approximately the same water levels. These are Baro to Onitsha, Lau and Makurdi; Itu to Onitsha, Lau, Makurdi and Obubra. Furthermore, Shintaku and Onitsha and Makurdi; Onitsha and Yola, Lau and Makurdi, as well as Makurdi and Obubra have approximately the same water levels (Fig. 4).

7 Construction of Infrastructures and Facilities

The FGN with a vision to develop the nation's IW embarked on the construction of ports, jetties and ramps/slipways across waterways in the country. Ports and jetties are important facilities that enhance the functions of the ships. Along the River Niger, the Onitsha port was constructed, commissioned and operational. The channelisation of Oguta Lake in Imo State and Excravos River in Delta State is in progress. Despite the huge amount expended on some of the projects, they are still uncompleted. As a result, there are worries regarding the appalling state of infrastructure services and facilities to enable NIWA to work at its best for improved national development in Nigeria (Fig. 5).

Table 3 Records of monthly average water levels in selected land Waterway Stations, 2021

Month	Baro (m)	Itu (m)	Shintaku (m)	Onitsha (m)	Yola (m)	Lau (m)	Makurdi (m)	Obubra (m)	Mean	Ranking
January	2.7	1.3	1.9	3.3	2.4	3.5	4	2	**2.6**	**10th**
February	2.7	1.1	1.7	3.3	2.2	3.6	3.7	2	**2.5**	**12th**
March	2.6	1.1	1.8	3.2	2.2	3.9	3.6	2.1	**2.6**	**9th**
April	2.3	1.3	1.7	3	2.2	3.7	3.7	2.3	**2.5**	**11th**
May	2	1.5	1.8	2.9	2.2	3.5	4.1	3.1	**2.6**	**8th**
June	2	2.1	2.6	4.1	2.3	3.6	5.1	4.4	**3.3**	**7th**
July	2.9	3.5	3.8	5.6	3.2	5	6.3	5.5	**4.5**	**4th**
August	4.4	4.2	5.7	7.9	4.6	6.9	7.6	6.5	**6.0**	**3rd**
September	6	5	7.6	10	5.2	7.8	9.3	7.4	**7.3**	**1st**
October	5.5	4.8	7.4	10.1	4.4	6.3	9	6.6	**6.8**	**2nd**
November	3.4	2.9	4.1	6.4	2.8	4.5	6	4.3	**4.3**	**5th**
December	2.7	1.7	2.2	3.9	2.4	3.6	4.6	2.8	**3.0**	**6th**
Mean	**3.3**	**2.5**	**3.5**	**5.3**	**3.0**	**4.7**	**5.6**	**4.1**		
Ranking	**6th**	**8th**	**5th**	**2nd**	**7th**	**3rd**	**1st**	**4th**		

Source NIWA (2022)

Fig. 3 River Niger at Onitsha. *Source* Authors' field survey, 2021

Fig. 4 Completed Onitsha Inland Port. *Source* Authors' field survey, 2021

8 Impact of Inland Water Transportation in Nigeria

8.1 Axle Load Reduction on Roads

The nation's road transportation system has come dangerously close to collapse, especially in Lagos, as a result of our continued reliance on trucks and trailers to deliver cargo from the Lagos Ports. When IWT is improved, Nigeria's highways will see some stability. Axle and destructive loads on the roads would be lowered because to the water's larger carrying capacity, which would lower the expense of road repair

Fig. 5 Baro Inland Port. *Source* Authors' field survey, 2021

nationwide. Additionally, the ferry services in Lagos from Mile 2 to Apapa to CMS reduce traffic gridlock on road as well as excessive loading of the road. Furthermore, an effective IWT system would greatly assist in reducing the carnage and pressure on the nation's roads.

8.2 Revenue Generation

Yakubu (2007) posited that revenue generated from IWT amounting to over N1.890 billion between 1998 and 2006, could be more if the IW sector was fully utilised. He asserted that more revenue could be generated through the improvement in dredging and development of IW river ports and the building of dockyards which in turn could attract more investors. According to the NIWA mission, the Nigerian Ports Authority (NPA) could generate money by charging registration fees for ships, customs duties, port fees and tariffs for the use of its facilities by ships that berth at sea. Revenue generation could be accomplished in accordance with the NIWA mandate through fees for ship registration, customs duties, port charges and tariffs realised by the Nigerian Ports Authority (NPA) for the use of its facilities by the vessels that berth at sea. In 2014 alone, NIWA had an Internally Generated Revenue (IGR) of N2,239,339,686 only.

Despite the low level of the development status of IWT in Nigeria, NIWA experienced a steady increase in IGR over the years which is an indication that more funds could be generated if IWT in Nigeria is well-developed and effectively utilised especially the River Niger, thus improving its contribution to the Gross Domestic Product (GDP) of Nigeria.

8.3 Promotion of Tourism

Tourism attraction equally contributes to revenue generation in Nigeria. The beaches in Lagos, the tourist beach in Port Harcourt and the Ibeano beach in Akwa Ibom attracted about 8.5 million people within and outside Nigeria (NTC, 2016). According to Fidelis (2017), Inland Waterways by 2020 would attract over 12 million tourists and possibly generate about $12.5 billion annually which could be channelled to developmental projects. Thus, efforts in line with improving tourism centres around the inland ports, especially at Rivers Niger and Benue would generate revenue for enhanced economic and national development.

8.4 Creation of Employment and Job Opportunities

Employment opportunities in IWT range from ship/boat building, and its repairs requiring expertise such as captains; engineers, stewards, etc. Igbokwe (2013) concluded that maintaining the industry in both the private and public sectors would provide about 10% of job opportunities in Nigeria. For instance, indirect employment in the riverine and coastal communities like Lagos and Bayelsa states have a workforce of about 9,000 and 5,500 personnel, respectively. These include boat operators, jetty workers and repairers, among others.

Table 4 and Fig. 4 indicate that the demand for barges and their carriage capacity is increasing from 2012 to 2021, which shows an increasing need for more investment for the provision of more vessels into the Nigerian Inland Waterways' transportation. Vessels that can carry dry bulk cargo are more required, followed by container carrying vessels and barges that carry liquid products.

Thousands of indirect jobs are generated by the various activities in the IW sector which would improve the living standard of the citizenry. This underscores the importance of an all-year-round navigable River Niger for viable IWT from Lagos, Warri, Onitsha and Baro in Niger State which would greatly increase job opportunities.

8.5 Industrial Growth and Development

IWT is significant as it promotes industrial growth and development. Onitsha being a popular market city and positioned along the River Niger's bank witnesses several tonnes of goods transported daily from Lagos by road. This has led to much pressure on the road as earlier highlighted. According to Etuk (2015), to cut down on transportation expenses, business owners choose to construct factories, industries and warehouses close to river ports. Presently, Nigeria's industrial base is very poor making it difficult for IWT to thrive like her counterparts in European countries. A viable IWT would improve the haulage of bulk cargo and raw materials for the

Table 4 Data on the number and capacity of barges required for Inland Waterways Transportation 2012–2021

Year	Container barges				Dry and break bulk barges				Refined oil product barges			
	No	Fleet capacity (tonnes)	% No	% Capacity	No	Fleet capacity (tonnes)	% No	% Capacity	No	Fleet capacity (tonnes)	% No	% Capacity
2012	12	6	**26.1**	**15.4**	26	18	**56.5**	**46.2**	8	15	**17.4**	**38.5**
2013	20	10	**20.0**	**12.5**	70	50	**70.0**	**62.5**	10	20	**10.0**	**25.0**
2014	21	12	**23.9**	**12.6**	56	61	**63.6**	**64.2**	11	22	**12.5**	**23.2**
2015	25	18	**29.8**	**15.9**	44	70	**52.4**	**61.9**	15	25	**17.9**	**22.1**
2016	30	22	**31.3**	**18.3**	48	72	**50.0**	**60.0**	18	26	**18.8**	**21.7**
2017	34	37	**31.5**	**25.7**	52	79	**48.1**	**54.9**	22	28	**20.4**	**19.4**
2018	40	40	**33.3**	**26.7**	55	80	**45.8**	**53.3**	25	30	**20.8**	**20.0**
2019	41	56	**35.3**	**32.7**	50	85	**43.1**	**49.7**	25	30	**21.6**	**17.5**
2020	45	75	**42.1**	**37.1**	36	95	**33.6**	**47.0**	26	32	**24.3**	**15.8**
2021	50	100	**45.5**	**41.7**	30	100	**27.3**	**41.7**	30	40	**27.3**	**16.7**
Mean	**31.8**	**37.6**			**46.7**	**71**			**19.0**	**26.8**		
Ranking	**2nd**				**1st**				**3rd**			

Source NIWA (2022)

Ajaokuta steel plant and the Itakpe iron ore processing plant from the seaports. This would replace the road that could transport 496 million tonnes of raw materials and approximately 2 million tonnes of finished goods.

8.6 Sustainable Development

Water transportation with a far less energy requirement remains the most environmentally friendly mode of transportation thus promoting sustainable transportation. Environmental degradation, lower pollution and depletion of natural resources are least experienced with water transportation. IWT is also cheap requiring lower operating costs as there is no wear and tear on the waterways demanding perennial high maintenance costs. Additionally, bulky and heavier goods are transported over longer distances at the cheapest comparative economic cost.

8.7 Sustainability of Nigerian Inland Waterways System

For thousands of enterprises throughout Nigeria, the Inland Waterways system may provide affordable, dependable and sustainable transportation. By removing major physical, technological or bureaucratic bottlenecks, Inland Waterway transport helps to divert traffic from roads. It will become even more effective by providing freight forwarders with a high-quality service that other land modes find it difficult to compete with in terms of price and dependability. Additionally, due to their efficiency and cleanliness, the usage of Inland Waterways will help the environment. The shipping sector and freight forwarders have the capacity and opportunities to take advantage of and win new business for Inland Waterway transportation.

However, Inland Waterways' substantial capacity alone is insufficient to boost their market share and modal split in comparison to road and rail transportation. The IWT industry must better integrate into seamless door-to-door transport chains, including effective transshipment operations and terminal hauls, in order to capture and maintain market share in growing markets and market niches, such as those for biomass, containers, bulky and heavy goods or waste and recycling materials. In addition, the IWT industry must comply with the increasingly sophisticated needs and requirements of supply chain and distribution managers. It is crucial to strengthen the connection between IWT and maritime shipping in order to accomplish the twin goals of expanding into new markets and improving intermodal transport and logistics chains. Promoting and regulating the usage of river-sea vessels—inland ships operating international runs between river and sea ports of different nations and coastal runs between river and sea ports of the same country—would be one way to accomplish this goal and relieve the congestion in maritime traffic. Getting rid of the wait times for loading/unloading commodities from inland vessels at the seaports is another significant problem in this sector.

To increase the appeal of the profession and deepen staff training, adequate transport and logistics policies are required. To guarantee complete traceability of commodities for their clients, IWT operators regularly modernise and expand their boats, create new transshipment strategies, establish regular container transport lines and increase their usage of information technologies.

Carbon emissions and global warming have emerged as crucial challenges for the future of IWT. First, whenever practicable, a modal shift away from road transportation can help to reduce the carbon emissions of the transportation industry. However, in order to keep this advantage, efforts must be made to guarantee that IWT advancements keep pace with the ongoing decrease in CO_2/t-km (CO_2 intensity) in road transport. There is a chance that significant fluctuations and decreased water depths will have an effect on IWT. There are possibilities where rivers will experience very little impact from climate change. All potential outcomes need to be taken into account when analysing the potential effects on inland navigation. IWT must simultaneously endeavour to retain and grow its competitive edge in environmental friendliness through research and innovation, taking into account, for example, the usage of alternative fuels.

Although no significant changes to the institutional framework of inland navigation in Nigeria are anticipated, governments and other stakeholders must have ongoing, inclusive consultation and coordination mechanisms in order to coordinate their policies and regulations and further harmonise rules and legal frameworks for an effective and long-lasting IWT.

Recent studies of the IWT labour market, for example, revealed that more workers will be needed in all areas covered by inland navigation due to the current technical developments, including investments in larger vessels and vessels with new capacity as well as the likely long-term increase in transport volume. In addition, a sizeable portion of the labour force is anticipated to quit the industry over the next ten to twenty years due to the age composition of the current IWT workforce.

9 Conclusion

This research work appraised Nigeria's inland water transportation system as it affects national development (ND). The research discovered that there are numerous challenges affecting the use of the river for IWT. Prominent among the challenges is the issue of the dredging of the River Niger which an in-depth study revealed that a lack of political will and commitment had deterred a capital dredging of the lower River Niger. An overview of the inland water transportation system from the colonial era to the post-independence era also revealed that IWT was the most viable mode of transportation during the colonial era and produced the revenue that was used to construct the rail and road infrastructures which affected the use of IWT. But the discovery of oil and subsequently the oil boom of the 1970s further affected the utilisation of IWT due to the shift from dependence on agriculture to oil as a mainstay.

The study identified some contributions of IWT to national and sustainable development to include, revenue generation, creation of employment and promotion of industrial growth and development. Others are the provision of safe and cheaper means of transportation, reduction of pressure on the roads as well as promotion of tourism. However, the challenges linked to the use of the River Niger for IWT include an ineffective legal framework, lack of political will and commitment on the part of FGN, inadequate skilled manpower and the impact of climate change. Others include the nature of the River Niger, the poor state of infrastructures, insufficient funding and insecurity among others.

Despite the challenges, there are significant plans to improve IWT using the River Niger by the FGN such as the construction of the Warri Port. This is however in a bid to reduce the traffic gridlock experienced in Lagos due to the congestion at the Apapa port. The study also brought to the fore the yearnings of the traders union in Onitsha and the environment in dire need of the construction of an inland port in Onitsha as well as the dredging of the River Niger to make it navigable all year round.

The execution of capital and periodic dredging of the River Niger and the inclusion of the private sector/foreign investors in infrastructural development will mitigate numerous challenges. While the periodic dredging of the River Niger will make the River navigable all year round. The inclusion of the private sector/foreign investors in infrastructural development will mitigate the challenges of the poor state of infrastructures through rehabilitation of the infrastructures and completion of abandoned projects. This could be achieved by the FGN in partnership with the PPP in repairing dockyards, water channels, barges, jetties, water clearing machines and boats by the end of 2022. Additionally, the establishment of an Inland Water Intervention Fund will mitigate the challenge of insufficient funding which seeks to increase funding to complement budgetary allocation.

References

Abdulmalik, A. (2016, December 14). *An interview conducted with Mr A. Abdulmalik, a personal assistant to the Ambassador of Egypt in Nigeria on inland waterways and national development, Maitama Abuja.*

Abubakar, R. D. (2000, June 12). *Providing ways and means on Nigerian Waterways* (p. 6). NIWA Publication.

Aderamo, A. J., & Mogaji, S. A. (2010). Rural transportation of public facilities in Nigeria: A case study of Edu Local Government Area of Kwara State. *Journal of Human Ecology, 29*, 171. https://doi.org/10.1080/09709274.2010.11906260

Anyam, R. W. (2003). *The complexity and dynamics of man-environment interactions in Nigeria.* A paper presented at the first stakeholders' workshop on Environmental Policy for Benue State held on October 15, 2003 at Benue State University, Makurdi.

Badejo, D. (1997). Inland water transport development in Nigeria. In K. Adeniji & A. Adesanya (Eds.), *Challenges of transport planning and management in Nigeria NISER*, Ibadan. Chapter 12.

Badejo, B. (2010, October 13). *Developing inland waterways transportation in Nigeria: What role for the private Sector*. Being a lecture delivered at the Lagos State Ministry of Transportation, Alausa, Ikeja, p. 8.

Daniel, E. (2012). *The role of inland water transport in support of further regional integration*. http://nileis.nilebasin.org/system/files/Nile%20SoB%20Report%20Chapter%207%20-%20Transport.pdf. Accessed 26 October 2016.

Douglas, I. E. (2001). *Impact of navigation on socio-economic development of West and Central Africa sub-region and as a veritable tool for the promotion of tourism and sports*. A paper presented at a 3-day International Conference on Navigation activities within the Niger River Basin West and Central Africa Sub-region, held at Abuja on the 26th–28th June 2001.

Ezenwaji, E. E. (2010). *Constraints on effective water transportation in some riverine communities of old Anambra L.G.A in Anambra State*. A paper delivered at the 2010 Rural Development Seminar with the theme Rural Transportation in Nigeria at Imo State University, Owerri between March 31 and April 1, 2010.

Federal Ministry of Transportation National Inland Waterways Authority. (2010). *Inland waterways masterplan and bankable feasibility studies*.

Fellinda, L. (2006). World's water transport needs further development. *Journal for Transport and Development, 1*, 68–72.

Fidelis, A. (2017). *An interviewed on inland waterways and national development in Nigeria conducted with Mr Fedelis, Deputy General Manager Marine at NIWA Abuja Office on January 5, 2017*.

Igbokwe, M. I. (2013). *Major problems associated with the Nigerian ports system and suggested solutions*. An Unpublished Seminar Paper, Federal University of Technology Owerri.

Iloje, I. J. (1984). *Regional geography of Nigeria*. Macmillan Publisher.

Kimba, G. (2008). Development of inland waterways in Freetown, Sierra Leone. *Journal of African Integration, 3*, 91–100.

Kingsley, E. A. (2017, January 16). *Interviewed with Mr Kingsley, E.A. Director Research and Development, Federal Ministry of Transport on inland waterways and national development in Nigeria at the Central Business District Area, Garki, Abuja*.

Lawal, T., & Oluwatoyin, A. (2011). National development in Nigeria: Issues, challenges and prospects. *Journal of Public Administration and Policy Research, 3*(9), 238.

Nasir, H. (2014, April 23). *An interview conducted with Mr H Nasir, Manager of Research and Development NIWA*.

National Inland Waterways Authority. (2014). *Summary of hand-over notes to transition committee* (p. 23). Federal Ministry of Transport.

National Tourism Centre, Lagos, May 2016, p. 12.

Ndahi, A. A. (2017, January 5). *An interviewed on inland waterways and national development in Nigeria conducted with Mr A.A. Ndahi, Principal Statistics Officer II at NIWA Headquarters Kogi*.

Ndikom, O. B. C. (2008). *The Kernel concept of shipping operations policies and strategies: The industry overview* (p. 5). Bunmice Publishers.

Ndikom, O. B. C. (2013). A critical assessment of the inland waterways operations and management on the development of the Nigerian maritime industry. *Greener Journal of Environmental Management and Public Safety, 2*(2), 99–117.

Nigerian Inland Waterways Authority (NIWA). (2013). Annual Report Publication.

Nigerian Inland Waterways (NIWA). (2015). Annual Report.

Nigerian Inland Waterways Authority (NIWA). (2016). Annual Report Publication.

Ojile, M. O. (2006). *Draft report of the socio-economic characteristics for the idealization of the Port Harcourt-Warri Road*. Submitted to Messer Allots Nigeria. Limited on behalf of the Federal Ministry of Works, Abuja, p. 60.

Ovuworie, I. U. (2014, June). *Transportation and national development in Nigeria: An assessment of inland waterways*. A research project submitted to the National Defence College Nigeria in partial fulfilment of the requirement for the award of Fellow of the National Defence College (FDC).

Rahman, M. (1994). National transport system in Bangladesh. *Asian Transport Journal, 21*, 18–23.

Raphuram, G. (2004). Integrating coastal shipping with national transport network in India. *Proceedings of the International Association of Maritime Economics Annual Conference, 11*(1), 39–48.

Review of Maritime Transport. (2007). *United Nations Conference on Trade and Development New York and Geneva* (p. 12).

Salawa, A. (2016). *The role of inland water transport in support of further regional integration* (p. 12). Tawakal Printing Press.

Udo, R. K. (1970). *Geographical regions of Nigeria*. Oxford University Press.

Usman, U., & Oyesiku, K. (2015, May 31). *Transportation in Nigeria: The redemption options*. http://leadership.ng/news/metro/437384/transportation-in-nigeria-the-redemption-options. Accessed 3 November 2016.

World Bank Report. (2009, May). *Sustainable development of inland waterway transport in China* (p. 8).

Yar'Adua, A. A. (2016, December 16). *Former Managing Director NIWA, phone interview on Inland Waterways and National Development in Nigeria*.

Yeung, A. (2016, December 12). *An interviewed on inland waterways and national development in Nigeria conducted with Mr A. Yerung, Personal Adviser to the Ambassador of China in Nigeria at the Central Area Abuja*.

The manufacturer's authorised representative in the EU is Springer Nature Customer Service Centre GmbH, Europaplatz 3, 69115 Heidelberg, Germany. If you have any concerns regarding our products, please contact ProductSafety@springernature.com

Printed and bound by CPI Group (UK) Ltd, Croydon, CR0 4YY
01/10/2026
02225892-0001